T0176664

PHYSICS OF MAGNETIC NANOSTRUCTURES

PHYSICS OF MAGNETIC NANOSTRUCTURES

FRANK J. OWENS
Department of Physics
Hunter College and the Graduate School
City University of New York

For general information on our other products and services or for technical support, please contact our Customer Care Department within the United States at (800) 762-2974, outside the United States at (317) 572-3993 or fax (317) 572-4002.

Wiley also publishes its books in a variety of electronic formats. Some content that appears in print may not be available in electronic formats. For more information about Wiley products, visit our web site at www.wiley.com.

Library of Congress Cataloging-in-Publication Data:

Owens, Frank J.
Physics of magnetic nanostructures / Frank J. Owens.
    pages   cm
  Includes bibliographical references and index.
  ISBN 978-1-118-63996-2 (cloth)
1. Nanostructured materials.   2. Magnetic structure.   I. Title.
  TA418.9.N35O844 2015
  620.1′1597–dc23
                                                2015004419

Printed in the United States of America

10 9 8 7 6 5 4 3 2 1

1   2015

CONTENTS

PREFACE

When solids have dimensions of nanometers, the properties of the solids change. Such properties as strength, melting temperature, color, electrical conductivity, thermal conductivity, reactivity, and magnetic properties are affected, and the magnitude of the change depends on the size of the solid in the nanometer regime. The changes also depend on the number of dimensions that are nanometers in size. Thus, size in the nanometer regime can be used to design and engineer materials with new and possibly technologically interesting properties. The development of applications based on nanoscience research is referred to as nanotechnology. Because of this potential, nanotechnology and nanoscience have generated much interest in recent years in the materials science, chemistry, physics, and engineering communities. As a result, chemistry, physics, materials science, and engineering departments at universities are developing courses in the various subfields of nanotechnology.

Nanostructured magnetic materials have a particularly strong possibility for developing new and interesting applications. In fact, there are already a number of technologies that employ nanostructured magnetic materials such as computer data storage and ferrofluids. The development of digital computers capable of handling large software programs has created a strong need for increased storage capability. Storage density has almost doubled every year. This is a result of a major research thrust in developing magnetic nanoparticles of smaller sizes with appropriate properties, which enables increased storage density. This increased density of smaller magnetic nanoparticles has driven the need to develop more sensitive methods of reading the storage devices because smaller particles have lower magnetizations. Nanosized magnetoresistive materials such as magnetic tunnel junctions have the potential to increase the sensitivity of reading devices.

Presently, there is also active research to develop new applications in the medical field. Such ideas as using magnetic nanoparticles for targeted delivery of drugs and enhancement of images in magnetic resonance imaging are presently under investigation. Using magnetic nanoparticles loaded with chemotherapeutic drugs to deliver them directly to the tumor has a large potential to eliminate the negative side effects of the drugs. Understanding this research and its possibilities requires knowledge of the basic ideas and properties of magnetic materials, how they are measured, and how nanosizing affects these properties.

This book, *The Physics of Magnetic Nanostructures*, is intended to provide this understanding as an introduction to the subject for those who wish to learn about the field or become involved in research on the subject. With omission of some sections, the book could also be the basis of a senior undergraduate or graduate level textbook on how and why reducing the size of solids to nanodimensions changes magnetic properties. Thus, exercises have been included at the end of each chapter. The objectives of the book are to describe how magnetic properties depend on the size and dimension in the nanometer regime and to explain using relatively simple models of the solid state why these changes occur. Experimental methods for measuring the magnetic properties are described as the data from them are first presented.

The first chapter presents a basic overview of the effect reducing the size of a solid to nanometers on the fundamental properties of the materials. The next chapter reviews the physics of magnetism and methods of measuring magnetic properties, which is necessary to understand how nanosizing affects magnetism. The remaining chapters discuss various kinds of magnetic structures and how nanosizing influences their magnetic properties. This includes two chapters that present potential and actual applications, one on devices and the other on medical applications.

ACKNOWLEDGMENT

The author would like to thank Prof. Emeritus Charles Poole of the University of South Carolina who taught me much about the art of writing science books.

1

PROPERTIES OF NANOSTRUCTURES

Nanostructures are generally considered to consist of a number of atoms or molecules bonded together in a cluster with at least one dimension less than 100 nm. A nanometer is 10^{-9} m or 10 Å. Spherical particles having a radius of about 1000 Å or less can be considered to be nanoparticles. If one dimension is reduced to the nano range, while the other two dimensions remain large, then we obtain a structure known as a well. If two dimensions are reduced, while one remains large, the resulting structure is referred to as a wire. The limiting case of this process of size reduction in which all three dimensions reach the low nanometer range is called a dot. Figure 1.1 illustrates the structures of rectangular wells, wires, and dots. This chapter will discuss how the important properties of materials such as the cohesive energy and the electronic and vibrational structure are affected when materials have at least one length in the nanometer range. Elementary models of the solid state will be used to explain why the changes occur on nanosizing.

1.1 COHESIVE ENERGY

The atoms or ions of a solid are held together by interactions between them, which can be electrostatic and/or covalent. The electrostatic interaction is described by the Coulomb potential between charged particles. Covalent bonding involves overlap of wave functions of outer electrons of nearest neighbor atoms in the lattice. A crystal is stable if the total energy of the lattice is less than the sum of the energies of the atoms or molecules that make up the crystal when they are isolated from each other. The energy difference is the cohesive energy of the solid. As materials approach nanometer dimensions, the percentage of atoms on the surface increases. Figure 1.2 demonstrates a plot of the percentage of atoms on the surface of a hypothetical face-centered cubic (fcc) structure having a lattice parameter of 4 Å. Appendix A provides

Physics of Magnetic Nanostructures, First Edition. Frank J. Owens.
© 2015 John Wiley & Sons, Inc. Published 2015 by John Wiley & Sons, Inc.

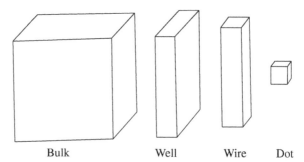

FIGURE 1.1   Structures corresponding to a rectangular well, wire, and dot having one, two, and three dimensions of nanometer length, respectively.

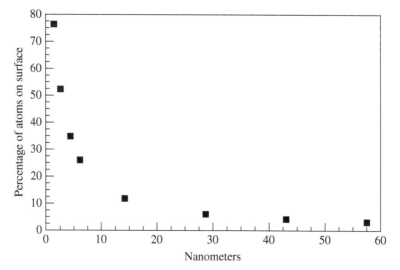

FIGURE 1.2   Percentage of atoms on surface of a face-centered cubic lattice versus particle size. The lattice parameter is 4 Å.

a table relating the diameter of spherical nanoparticles to the number of atoms in the particle and the percentage on the surface. Below about 14 nm, more than 10% of the atoms are on the surface. This holds true for metallic particles as well as ionically and covalently bonded materials. Since the atoms on the surface have less nearest neighbor atoms, this means that the cohesive energy of an ionic solid decreases as the size is reduced in the nanometer range. One of the results of this decrease in cohesive energy is an increase in the separation of the constituents of the lattice. Figure 1.3 shows an X-ray diffraction measurement of the lattice parameter of the ionic solid CeO_2 as a function of particle size showing the increase in the lattice parameter as the particle size is reduced. This results in a reduction of the strength of the interaction between the ions of the solid and thus a reduction in the cohesive energy.

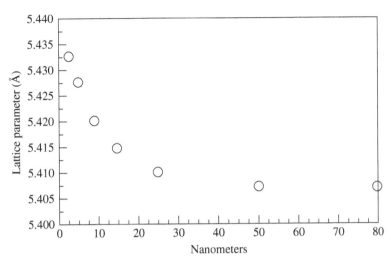

FIGURE 1.3 Experimental measurement of the lattice parameter of the ionic crystal, CeO_2, versus particle size. (Reproduced with permission from Zhang et al. [1]. © 2002, AIP Publishing LLC.)

Ionic solids are ordered arrays of positive and negative ions such as sodium chloride, which is an fcc structure of positive sodium ions and an interpenetration of fcc negative chlorine ions. The interaction potential between the ions of charge Q is electrostatic, $\pm Q/r$. Ions of opposite sign are attracted, while ions of the same sign repel each other. The total electrostatic energy of any one ion i is U_i given by the sum of all the Coulomb interactions between the ith ion and all positive and negative ions of the lattice:

$$U_i = \sum_j U_{ij} = \sum_j \pm \frac{Q^2}{r_{ij}} \qquad (1.1)$$

where r_{ij} is the distance between ions i and j. In the case of the interaction between the nearest neighbors, a term has to be added to Equation 1.1 to take into account that the electron core around the nucleus repels those of the nearest neighbors. The form of this has been derived from experiment and is given by

$$\lambda \exp\left(\frac{-r_{ij}}{\rho}\right) \qquad (1.2)$$

The constants λ and ρ for NaCl are 1.75×10^{-9} ergs and 0.321. The larger the r_{ij}, the smaller the cohesive energy. Thus, for ionic crystals, the cohesive energy decreases as the lattice parameter increases. Figure 1.4 shows a plot of the experimentally determined cohesive energy of crystals having the NaCl structure versus lattice parameter. The reduction of the cohesive energy also affects other properties such as

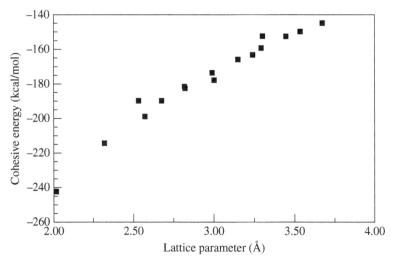

FIGURE 1.4   Cohesive energy of ionic alkali halide crystal having NaCl structure versus lattice parameter.

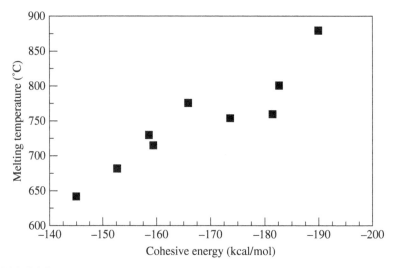

FIGURE 1.5   Melting temperature of some alkali halide crystals versus cohesive energy.

the melting temperature. Figure 1.5 illustrates a plot of the melting temperature of some alkali halides versus cohesive energy. In general, when materials have nanometer dimensions, the melting temperature decreases.

Most of the experimental observations of the effect of size of metal nanoparticles on the lattice parameters show it decreases as the diameter decreases. The decrease is attributed to the effect of surface stress. The surface stress causes small particles to be in a state of compression where the internal pressure is inversely proportional to the radius of

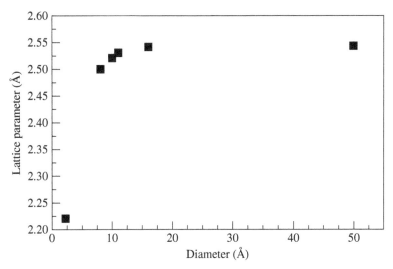

FIGURE 1.6 Measured lattice parameter of copper versus particle size. (Reprinted with permission from Ref. [2]. © 1986 by the American Physical Society.)

the particle. Figure 1.6 demonstrates a plot of the measured decrease in the lattice parameter of copper versus the diameter in angstroms. Notice that the changes don't occur until the diameter reaches a very small value of 0.9 nm. In the case of gold, measurements show that at 3.5 nm the lattice parameter has decreased to 0.36% of the bulk value. In aluminum, significant changes are not observed until the particle size is below 1.8 nm. In the discussion of models of the electronic properties of metals in the following section, it will be assumed that the lattice parameter is not significantly dependent on particle sizes for values greater than 4 nm. As will be seen, the number of atoms in a metal nanoparticle has a much more significant influence on the electronic structure.

Metals conduct electricity because the outer electrons of the atoms of the solid are delocalized and hence free to move about the lattice. This makes the development of a theory of binding energy of metals a bit more complex than for ionic or covalent solids. Because the outer electrons of the atoms can be itinerant, the atoms can be considered to be positively charged. The binding energy of a metal can be treated as arising from the Coulomb interaction of a lattice of positive ions embedded in a sea of negative conduction electrons. One relatively simple model is to consider the binding energy to be the interaction of a positive point charge e with a negative charge −e distributed uniformly over a sphere of radius R_0 and volume V_0 equal to the atomic volume. The cohesive energy on this simple model can be shown to be [3]

$$U_c = -\frac{0.9e^2}{R_0} + \left(\frac{3}{5}\right)[E_f] \tag{1.3}$$

where E_f is the Fermi level, the top occupied energy level, of the metal nanostructure. One would not expect the atomic volume to change significantly with reduced

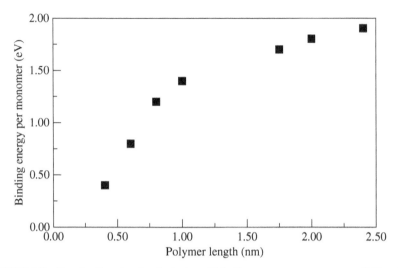

FIGURE 1.7   Density functional calculation of binding energy per monomer, C_2H_2, of the linear chain polymer, polyethylene versus polymer length. (Adapted from Ref. [4].)

size, and thus, the first term in Equation 1.3 will not change much with nanosizing. In the next section, it will be shown that the magnitude of the Fermi energy E_f increases as the particle size is reduced. From Equation 1.3, this implies that the cohesive energy of metals decreases as the particle size decreases in the nanometer regime. However, this decrease will not happen until the particle size is quite small.

Solids such as silicon are covalently bonded, meaning that the bonding involves overlap of the wave functions of nearest neighbors. Reduction in particle size would lead to a decrease in the binding energy. Figure 1.7 shows a plot of the calculated binding energy per monomer as a function of the length of a polyacetylene polymer in the nanometer range. Polyacetylene is a chain of covalently bonded monomers, C_2H_2. The binding energy of the monomer in the chain is given by

$$BE = BE_{pa} - nBE_{monomer} \qquad (1.4)$$

where BE_{pa} is the binding energy of the polymer, $BE_{monomer}$ is the binding energy of the monomer, and n is the number of monomers in the chain.

Generally, the interaction between the constituents of organic crystals is a dipole–dipole potential, which depends on their separation as $1/R^3$. Because of its short range, the crystals would have to be quite small to affect the lattice parameter. The potential describing the interaction between atoms in inert gas solids is the Lennard–Jones potential, which has a $1/R^6$ dependence on the atomic separation, meaning that the reduction in the size of the crystal to nanometers will have little effect on the cohesive energy. However, there have been no experimental studies of the effect of size on the lattice parameters of organic or inert gas solids.

1.2 ELECTRONIC PROPERTIES

One of the simplest models of the electronic structure of metals treats the conduction electrons as though they see no potential at all, but are confined to the volume of the solid. The model is best applicable to monovalent metals such as lithium, sodium, or potassium, where the ion cores only occupy about 15% of the volume of the solid. The energies are obtained by solving the Schrödinger wave equation for $V(r)=0$, with boundary conditions. For the case of a one-dimensional system, the wave equation has the form

$$-\left[\frac{h^2}{2m}\right]\frac{d^2\Psi_n}{dx^2} = E_n\Psi_n \tag{1.5}$$

where Ψ_n is the wave function of the electron in the nth state and E_n is the energy. The boundary conditions for a one-dimensional lattice of length L are

$$\Psi_n(0) = 0 \text{ and } \Psi_n(L) = 0 \tag{1.6}$$

The eigenvalues obtained by solving Equation 1.5 are [3]

$$E_n = \left[\frac{h^2}{2m}\right]\left[\frac{n}{2L}\right]^2 \tag{1.7}$$

where n is a quantum number having integer values 0, 1, 2, 3,..., etc.

Equation 1.7 is useful in understanding how the electronic structure of metals is affected when the dimensions are nanometers. The separation between the energy levels of state n and $n+1$ is

$$E_{n+1} - E_n = \left[\frac{h^2}{8mL^2}\right][1+2n] \tag{1.8}$$

It is seen from Equation 1.8 that as the length of the chain, L, decreases, the separation between energy levels increases, and eventually, the band structure opens up into a set of discrete levels. This also means that the density of states, the number of energy levels per interval of energy, will decrease with size. When the energy levels are filled with electrons, only two electrons are allowed in each level because of the Pauli exclusion principle. These two electrons must have different spin quantum numbers m_s of $+1/2$ and $-1/2$, meaning that the two electron spins in each level n are antiparallel and there is no net spin in the level. The Fermi energy is the energy of the top filled level, which for a monovalent metal will have the quantum number $n_f = N/2$ where N is the number of atoms in the solid. Thus, for the one-dimensional solid, the Fermi energy is obtained from Equation 1.7 to be

$$E_f = \left[\frac{h^2}{2m}\right]\left[\frac{N}{4L}\right]^2 \tag{1.9}$$

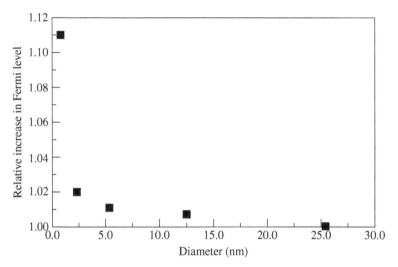

FIGURE 1.8   Calculation of the relative shift of the Fermi level for copper versus particle diameter.

In a manner analogous to the derivation of Equation 1.9 for the Fermi level in one dimension, the Fermi level in the free electron model of metals in three dimensions is obtained to be

$$E_f = \left[\frac{h^2}{2m(\pi)^2} \right] \left[\frac{3\pi^2 N}{V} \right]^{2/3} \tag{1.10}$$

Figure 1.8 demonstrates a calculation of the relative change of the Fermi level for the face-centered copper lattice versus diameter. The plot shows that the Fermi level does not increase significantly until the particle size is quite small, less than 2.5 nm for copper.

An important property of solids is the density of states, that is, the number of energy levels per interval of energy given by $D(E) = dN/dE$. The density of states determines a number of properties of solids such as the electronic specific heat and the magnetic susceptibility arising from the conduction electrons. The density of states depends on the dimensionality of the material. For a wire, a one-dimensional material, the density of states can be calculated from Equation 1.7 and has the form $1/2\,C_1 E^{-1/2}$. For two dimensions, a well, the density of states can be shown to be $D(E) = C$, and for three dimensions, $D(E) = C_3 E^{1/2}$. Thus, we see that the density of states of a material depends on its dimensions. Figure 1.9 gives a plot of the density states for the different dimensions.

The band gap of a material is the energy separation between the top filled energy level and the first unfilled level. One way to measure the effect of nanosizing on the band gap of a nanosized semiconductor is to measure the absorption of light as a function of the wavelength for different particle sizes. An absorption will occur when the energy of the photon of the light is equal to or greater than the band gap. When this happens, an electron is excited from the valence band to the conduction band and

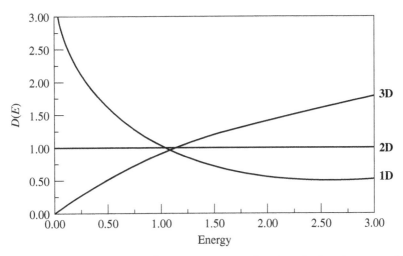

FIGURE 1.9 Plot of the density of states versus energy based on the free electron model of metals for structures having one, two, and three dimensions of nanometer length.

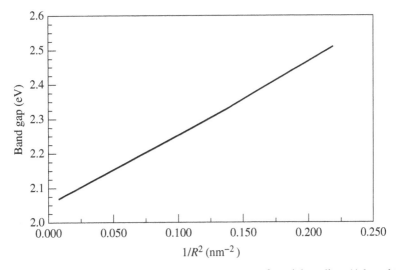

FIGURE 1.10 Band gap of CdSe versus inverse square of particle radius. (Adapted from Ref. [5].)

light energy is absorbed. Nanosized cadmium selenide is a semiconductor. Figure 1.10 illustrates a plot of the measured band gap of cadmium selenide versus the inverse square of the particle radius showing a nearly straight line dependence. A nanowire is a nanostructure in which two dimensions have nanometer size, and the other is large, generally greater than microns. Figure 1.11 shows the result of the measurement of the band gap of a silicon nanowire versus its diameter, showing the band gap increasing as the diameter of the wire is reduced.

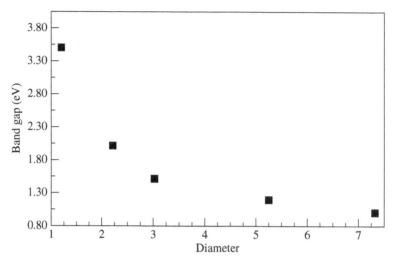

FIGURE 1.11 Measured band gap of silicon nanowire versus diameter in nanometers (Adapted from Ref. [6].)

1.3   QUANTUM DOTS

Semiconductors are materials having small band gaps typically ranging from 2.5 to 0.18 eV such that an electron at higher temperatures can be excited from the valence band to the conduction band providing a source of charge carriers for current. For example, silicon, which is widely used in transistor devices, has four valence electrons that are shared in covalent bonds with four neighboring silicon atoms in the lattice. The electronic structure of the lattice is determined by this covalent bonding. Thus, Equations 1.6, 1.7, 1.8, 1.9, and 1.10 cannot be used to describe the electronic structure of bulk semiconductors. There is one situation where similar equations can be used to describe the energy levels. This is at the size where quantum confinement occurs. As the dimensions of wires, wells, and dots are decreased, there is a size where the separation of the surfaces of the particles is in the order of the wavelength of the charge carriers. The charge carriers are said to be confined, and the effect is referred to as the quantum size effect. At these sizes, the structures in Figure 1.1 are referred to as quantum wires, wells, and dots. At this size, the energy levels of the structures are not determined by the chemical nature of the atoms of the material but by the dimensions of the structure. It is interesting to note that the quantum size effect occurs in semiconductors at larger sizes because of the longer wavelength of conduction electrons and holes in semiconductors due to the larger effective mass. In a semiconductor, the wavelength can be as long as 1 μm, whereas in a metal, it is in the order of 0.5 nm.

A simple model that exhibits the principal characteristics of such a potential well is a square well in which the boundary is very sharp or abrupt. Square wells can exist in one, two, three, and higher dimensions, and for simplicity, a one-dimensional case will be considered.

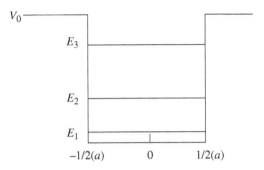

FIGURE 1.12 Energy levels of a finite potential well.

Standard quantum mechanical texts show that for an infinitely deep square potential well of width, a, in one dimension, the coordinate x has the range of values $-\frac{1}{2}(a) \leq x \leq \frac{1}{2}(a)$ inside the well, and the energies there are given by the expressions

$$E_n = \frac{h^2}{2m}\left(\frac{n^2}{a^2}\right)$$ (1.11a)

$$= E_0 n^2$$ (1.11b)

where $E_0 = h^2/2ma^2$ is the ground state energy and the quantum number n assumes the values $n = 1, 2, 3,\ldots$ The electrons that are present fill up the energy levels starting from the bottom, until all available electrons are in place. An infinite square well has an infinite number of energy levels, with ever-widening spacings as the quantum number n increases. If the well is finite, then its quantized energies E_n are smaller than the corresponding infinite well energies, and there are only a limited number of them. Figure 1.12 illustrates the case for a finite well of potential depth $V_0 = 7E_0$, which only has three allowed energies. No matter how shallow the well, there is always at least one bound state E_0. For the case of a cubic quantum dot having edges of length, a, the energy levels will be

$$E_n = \left[\frac{h^2}{2ma^2}\right](n_x^2 + n_y^2 + n_z^2)$$ (1.12)

Quantum dots such as the cubic dot, having energy levels given by the above equation, have been developed into one of the major applications of nanotechnology. The quantum dot laser is used in CD players to read the groves on the disk. The separation between the levels in the dot can be chosen by the value of a in Equation 1.12. There is a value of a in which the separation of the energy levels from the conduction band can be in the infrared (IR) frequency range. This means an IR photon can excite an electron to the conduction band and application of a voltage produces a current. This is the basis for the use of the quantum dot as an IR detector. It is possible with appropriate excitation to produce a population inversion in the energy levels of the dot. This means that one of the upper levels has more electrons than a lower level, which is necessary to produce laser light.

1.4   VIBRATIONAL PROPERTIES

The constituents of a solid lattice vibrate. The specific frequencies, called the normal modes of vibration, are determined by the nature of the interaction between constituents of the lattice and the symmetry of the lattice. The vibrational frequencies of solids can be measured by IR spectroscopy and Raman spectroscopy. IR spectroscopy measures the absorption of IR light when it induces a transition from the $N=0$ vibrational state to the $N=1$ state. The basis of Raman spectroscopy is illustrated in Figure 1.13. Laser light is used to excite the lowest energy level of a vibration to some higher level. The higher level excited state then decays back to the lowest level. However, some of decay goes to a vibrational state above the ground state. The frequency of this emitted light is measured, and the difference between the frequencies of exciting laser light and the emitted light measures the vibrational frequency. When solids are reduced to nanometer dimensions, the frequencies generally decrease. Figure 1.14 shows a plot of the decrease in the frequency of the longitudinal optical mode of silicon as a function of particle size measured by Raman spectroscopy. There are two reasons for this decrease. It is seen earlier that the lattice parameter generally increases as the particle size decreases. This weakens the interaction between the constituents of the lattice and thus causes a lowering of the frequencies. The other reason for the decrease is phonon confinement, a process similar to electron confinement discussed earlier. This occurs when the dimensions of the solid are in the order of the wavelength of the lattice vibrations. The uncertainty principle can be used to explain phonon confinement.

The uncertainty principle says that the order of magnitude of the uncertainty in position ΔX times the order of magnitude of the uncertainty in momentum ΔP must at least be Planck's constant, h, divided by 2π, that is,

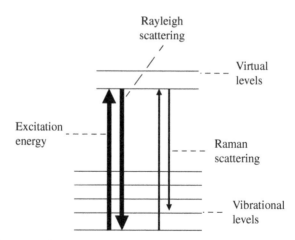

FIGURE 1.13   Illustration of excitation and emission of light from vibrational energy levels providing the basis of Raman spectroscopy.

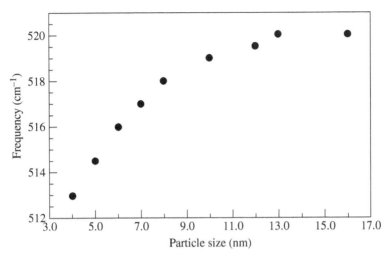

FIGURE 1.14 Plot of the frequency of the longitudinal optical mode of silicon versus the particle size measured by Raman spectroscopy. (Adapted from Ref. [7].)

$$\Delta X \Delta P \geq \frac{h}{2\pi} \qquad (1.13)$$

Let us assume that ΔX is the diameter of the nanoparticle, D, and that it can be measured accurately by some technique such as scanning electron microscopy. This means the uncertainty in the momentum P will have a range of values. It can be shown that the momentum of a phonon is $hk/2\pi$ where k is the wave vector given by ω/c and c is the velocity of light. Thus, we have $D\Delta k \geq 1$ or the minimum uncertainty in k is

$$\Delta k \sim \frac{1}{D} \qquad (1.14)$$

Raman spectroscopy measures frequencies at the center of the Brillouin zone, $k = 0$.

However, that is a precise value that Equation 1.14 shows has some uncertainty, meaning that there is a spread of values for k for a Raman measurement in small particles. This means nonzero values of k will contribute to the Raman spectrum. Let us assume that for small values of k, the dependence of the frequency on k, that is, the dispersion relationship, is quadratic:

$$\Delta \omega = ak^2 \qquad (1.15)$$

From this, it follows that

$$\Delta K = C \Delta \omega^{1/2} \qquad (1.16)$$

where C is a constant. Thus,

$$\Delta\omega = C\left[\frac{1}{D^2}\right] \tag{1.17}$$

However, if the dependence of the frequency on k is other than k^2, say k^γ, then the dependence of the frequency shift on particle diameter will be $1/D^\gamma$, which is what is observed with γ ranging from 1 to 1.5 depending on the material.

The vibrational density of states $D(\omega)$ is the number of vibrational modes per interval of frequency, $dN/d\omega$. For a one-dimensional line having N atoms, the number of vibrational modes is N. For a three-dimensional lattice, it is $3N$. The normal modes may be considered a set of independent oscillators with each oscillator having the energy ε_n given by

$$\varepsilon_n = \left(n + \frac{1}{2}\right)\left(\frac{h}{2\pi}\right)\omega \tag{1.18}$$

where n is a quantum number having integral values ranging from 0, 1, 2,... to n.

The average energy of a harmonic oscillator assuming a Maxwell–Boltzmann distribution function is

$$\langle \varepsilon \rangle = \frac{h\omega}{2\pi}\left[\frac{1}{2} + \frac{1}{\exp\left(\dfrac{h\omega}{\dfrac{2\pi}{k_B T}}\right) - 1}\right] \tag{1.19}$$

The total energy of a collection of oscillators is

$$E = \sum_k <\varepsilon_k>\left(\frac{h}{2\pi}\right)\omega_k \tag{1.20}$$

When there is a large number of atoms, the allowed frequencies are very close and can be treated as a continuous distribution allowing replacement of the summation by an integral, where $D(\omega)d\omega$ is the number of modes of vibration in the range ω to $\omega + d\omega$. It turns out it is more convenient to work in k space and deal with the number of modes $D(k)dk$ in the interval k to $k + dk$. As shown in Figure 1.15, the number of values of k in three dimensions between k and $k + dk$ will be proportional to the volume element on a sphere of radius k, which is given by

$$dV = 4\pi k^2 dk = D(k)dk \tag{1.21}$$

The dispersion relationship refers to the dependence of the frequency ω on the k vector.

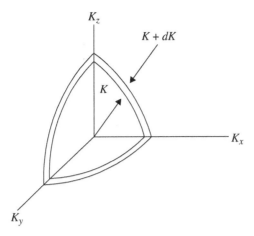

K_z

$K + dK$

K

K_x

K_y

FIGURE 1.15 A spherical shell in K space used to obtain the vibrational density of states for a three-dimensional solid.

One approximation due to Debye assumes that the relationship is linear, which is valid for low values of k, that is,

$$\omega(k) = uk \qquad (1.22)$$

From Equations 1.21 and 1.22,

$$D(\omega)d\omega = B\omega^2 d\omega \qquad (1.23)$$

where B is a constant, and thus,

$$D(\omega) = B\omega^2 \qquad (1.24)$$

In three dimensions, the total number of modes is $3N$, which means

$$\int_0^{\omega_d} D(\omega)d\omega = 3N \qquad (1.25)$$

where ω_d is the highest frequency that can propagate in the lattice and is referred to as the Debye frequency. Thus, for three dimensions in the Debye approximation, the density of states for $\omega < \omega_d$ is

$$D(\omega) = \frac{9N\omega^2}{\omega_d^3} \qquad (1.26)$$

In two dimensions, we would carry out the analogous derivation using a circle having area πK^2 obtaining

$$D(k)dk = 2\pi k dk \tag{1.27}$$

and

$$D(\omega)d\omega = \omega d\omega \tag{1.28}$$

$$\int_0^{\omega_d} D(\omega)d\omega = 2N \tag{1.29}$$

$$D(\omega) = \frac{4\omega N}{\omega_d^2} \tag{1.30}$$

Following the same procedure, the density of states in one dimension can be obtained to be

$$D(\omega) = \frac{N}{\omega_d} \tag{1.31}$$

Thus, the density of phonon states in the Debye approximation depends on the dimensionality of the material. Figure 1.16 shows a plot of phonon density of states versus the frequency for the different dimensions. Notice that the density of states decreases with the number of atoms N, which means the density of states will decrease in the nanometer range.

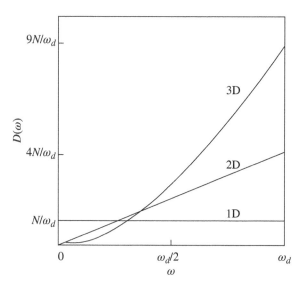

FIGURE 1.16 Density of phonon states versus frequency in the Debye approximation for one-, two-, and three-dimensional materials.

1.5 SUMMARY

This chapter has presented a discussion of how the important properties of solids, such as cohesive energy and electronic and vibrational structure, are changed when the dimensions of the materials are of nanometer length. Simple models of the solid state are used to explain why the effects occur. It is shown that the cohesive energy and vibrational frequencies decrease as the materials achieve nanometer sizes. Nanosizing causes the separation of the electronic energy levels of solids to increase, which causes a reduction in the density of states. It is shown that the effects depend on whether some or all of the dimensions are of nanometer length. The remainder of the book will deal with magnetic properties of nanostructures.

EXERCISES

1.1 At what particle size would you expect the melting temperature of CeO_2 to decrease? Explain your answer.

1.2 If the wavelength of an electron in a semiconductor is $0.75\,\mu m$, in order to create a quantum dot of this semiconductor, what volume would it have to have?

1.3 Suggest a method to make a tunable quantum dot laser.

1.4 If the dependence of the vibration frequency of a mode in a solid is $K^{1.5}$, obtain an expression for the dependence of the frequency of the mode on the particle size.

1.5 It is observed that when the particle size reaches a few nanometers, the line width of Raman spectra becomes broader. Why does this occur?

1.6 When the Raman spectra of Cu nanoparticles are measured, a line is observed that corresponds to that observed in CuO but occurs at a slightly lower frequency than in bulk CuO. Where is the CuO line coming from and why is at a lower frequency?

REFERENCES

1. F. Zhang et al. *Appl. Phys. Lett.* 80, 127 (2002).
2. P. A. Montano, et al. *Phys. Rev. Lett.* 56, 2076 (1986).
3. C. Kittel, *Introduction to Solid State Physics*, 3rd Ed. p. 201, John Wiley & Sons, Inc., New York, 1996.
4. F. J. Owens, *Physica* E25, 404 (2005).
5. A. I. Ekimov, *Solid State Commun.* 56, 921 (1985).
6. D. D. Ma et al. *Science* 299, 1874 (2003).
7. Z. Iqbal and S. Veprek, *J. Phys.* C15, 377 (1982).

2

THE PHYSICS OF MAGNETISM

2.1 KINDS OF MAGNETISM

Atoms whose energy levels are not totally filled have a net magnetic moment and in effect behave like small bar magnets. The value of the magnetic moment of a body is a measure of the strength of the magnetism that is present. Atoms in the various transition series of the periodic table have unfilled inner energy levels in which the spins of the electrons are unpaired, giving the atom a net magnetic moment. The iron atom has 26 electrons circulating about the nucleus. Eighteen of these electrons are in filled energy levels, which constitute the argon atom inner core of the electron configuration. The d-level of the $N = 3$ orbit contains only 6 of the possible 10 electrons that would fill it, so it is incomplete to the extent of 4 electrons. This incompletely filled electron d-shell causes the iron atom to have a strong magnetic moment.

When crystals such as bulk iron are formed from atoms having a net magnetic moment, a number of different situations can occur relating to how the magnetic moments of the individual atoms are aligned with respect to each other. Figure 2.1 illustrates some of the possible arrangements that can occur in two dimensions. The point of the arrow is the north pole of the bar magnet associated with the atom. If the magnetic moments are randomly arranged with respect to each other, as shown in Figure 2.1a, then the crystal has a zero net magnetic moment, and this is referred to as the paramagnetic state. The application of a dc magnetic field aligns some of the moments, giving the crystal a small net moment. In a ferromagnetic crystal, these moments all point in the same direction, as shown in Figure 2.1b, even when no dc magnetic field is applied. The whole crystal has a magnetic moment and behaves like a bar magnet producing a magnetic field outside of it. If a crystal is made of two types of atoms each having a magnetic moment of a different strength, they can also align in a parallel arrangement, and the crystal is called a ferrimagnet. Such a crystal will also have a net magnetic moment and behave like a bar magnet. In an antiferromagnet,

Physics of Magnetic Nanostructures, First Edition. Frank J. Owens.
© 2015 John Wiley & Sons, Inc. Published 2015 by John Wiley & Sons, Inc.

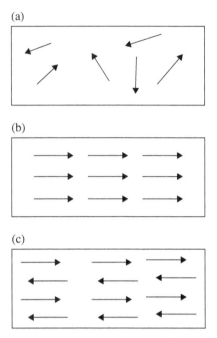

FIGURE 2.1   Illustration of various arrangements of individual atomic magnetic moments in two dimensions that constitute (a) paramagnetic, (b) ferromagnetic, and (c) antiferromagnetic materials.

the moments of identical atoms are arranged in an antiparallel scheme that is opposite to each other, as shown in Figure 2.1c, and hence, the material has no net magnetic moment. This book will be mainly concerned with ferromagnetic and to a lesser extent antiferromagnetic ordering in nanostructures.

2.2   PARAMAGNETISM

2.2.1   Theory of Paramagnetism

Consider a paramagnetic crystal such as $CuSO_4$ in which the copper has a spin ½. The magnetic moment of the Cu is given by

$$\mu = -g\left(\frac{eh}{2mc\pi}\right)\mathbf{S} \tag{2.1}$$

where g is a constant having a value close to 2, $\mathbf{S}$ is the spin vector, c is the velocity of light, and m is the mass of the electron. Since the crystal is paramagnetic, the magnetic moments of the Cu atoms are randomly oriented. This occurs because the interaction between the neighboring moments is weak and less than the lattice

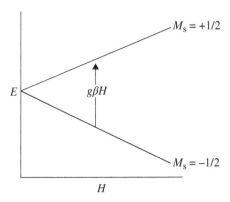

FIGURE 2.2   Illustration of the splitting of the m_s spin states of an unpaired electron in a dc magnetic field showing the radiation-induced absorption of energy that occurs when $h\nu = g\beta H$, which is the basis of the electron spin resonance measurement.

vibrational energy, which is in the order of kT. When a magnetic field H is applied, there is an interaction with the magnetic moment given by[1]

$$E = \mu \cdot H \tag{2.2}$$

The magnetic moments align parallel and antiparallel to the H field, and there are two energy levels labeled by the spin quantum numbers: $m_s = -1/2$ for the spin-down orientation and $m_s = 1/2$ for the spin-up orientation. The energy separation between the levels depends on the strength of the magnetic field H as shown in Figure 2.2. The population of the states depends on the temperature and the magnetic field strength. The fraction of the total number of paramagnetic atoms N in the spin-down state N_β and the spin-up state N_α can be obtained from Boltzmann statistics

$$\frac{N_\beta}{N} = \frac{e^x}{(e^x + e^{-x})} \quad \text{and} \quad \frac{N_\alpha}{N} = \frac{e^{-x}}{(e^x + e^{-x})} \tag{2.3}$$

where $x = \mu \cdot H / kT$. In order for the material to have a net magnetization, there must be a difference in the population between the two levels:

$$M = (N_\beta - N_\alpha)\mu = N\mu \left[\frac{e^x - e^{-x}}{e^x + e^{-x}} \right] = N\mu \tanh x \tag{2.4}$$

For $\mu \cdot H$ much less than kT, the $\tanh x \approx x$ and the magnetization becomes

$$M = N \left(\frac{\mu^2 H}{kT} \right) \tag{2.5}$$

[1]The magnetic fields B, H, and M and their units are defined in Appendix B.

The susceptibility, χ, defined as M/H, is then

$$\chi = \frac{N\mu^2}{kT} \tag{2.6}$$

Equation 2.6 is known as Curie's law and has been verified experimentally in many materials by showing that the susceptibility is linearly dependent on $1/T$ in the paramagnetic phase.

2.2.2 Methods of Measuring Susceptibility

AC susceptibility is a simple and widely used method to characterize the static and magnetic properties of materials. It allows a direct measurement of the susceptibility over a range of frequencies. Figure 2.3 shows a schematic of an ac susceptibility apparatus. An ac signal is applied to a primary coil P. Wound around the coil are two secondary coils S having an equal number of turns and thus equal inductance. One of the coils contains the sample to be characterized. The EMF induced in this coil is directly proportional to the magnetization of the sample. The difference of the AC voltage between the two coils is measured, which is proportional to the susceptibility.

Another method consists of an oscillator connected to an external coil. The sample is contained in the coil, which is in a cryogenic dewar between the poles of a magnet. The change in the frequency of the oscillator as a function of the strength of the applied dc magnetic field is proportional to the change in the real part of the susceptibility χ given by

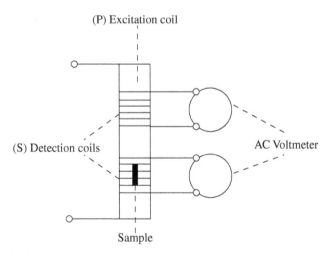

FIGURE 2.3 Schematic of apparatus to measure AC susceptibility consisting of two secondary coils having equal inductances wound around a primary coil. The sample is located in one of the secondary coils. The difference in voltage between the two coils is proportional to the susceptibility of the sample.

$$\frac{\Delta\omega}{\omega_0} = -2\pi F \chi \tag{2.7}$$

where F is the filling factor of the sample in the coil. Thus, a measure of the change in frequency measures the susceptibility. In principle, with a very stable oscillator and sensitive frequency meter, this can be a very sensitive method. The magnitude of the magnetization is obtained by comparing the measurements with frequency shifts of a sample of known magnetization. One drawback of this method is that the resonant frequency of the empty coil is temperature dependent because of the temperature dependence of the resistivity of the wire of the coil. Thus, temperature-dependent measurements have to be corrected for the temperature dependence of the frequency of the empty coil.

The most sensitive detector of magnetic properties is the superconducting quantum interference device (SQUID). It can detect magnetic fields as small as 10^{-11} T. It would be too far afield to go into the detailed working of a SQUID, so a qualitative overview of the principles behind it will be presented. It is based on the phenomena of superconductivity. Superconductivity is a state of matter where the resistance of a material is zero and the material is a perfect diamagnetic, meaning no magnetic field can penetrate the bulk of it. The detector uses a Josephson junction. The junction consists of a thin insulating material, about 10–20 Å thick, sandwiched between two superconducting metals. In many instances, the insulating layer is a thin oxide coating on an evaporated metal film. A voltage is applied to the junction, and it is then cooled below the transition temperature of the superconductor. When the voltage is turned off, it is observed that a dc current continues to flow as though the sandwich was one continuous slab of superconductor. In the superconducting state, the conduction electrons form electron pairs, called Cooper pairs, which have relatively long wavelengths, considerably longer than the interatomic spacing of the atoms of the metal. Also, the wave function of every pair is in phase with every other. The Cooper pairs are able to move through the insulating nonsuperconducting layer without breaking up. The phenomenon cannot be explained in classical terms, but requires the quantum mechanical representation of the pairs as waves. The explanation of how the current can cross the insulating layer involves the purely quantum mechanical phenomenon of tunneling. Normally, for an electron to pass from a conductor to an insulator, it must overcome an energy barrier that exists at the interface. Treating the electron as a classical particle, the only way the electron can get into the insulator is to possess an amount of energy greater than the energy associated with the barrier height. In quantum mechanics, the electron or electron pair is a probability wave. Quantum mechanics predicts that there is some probability that it can penetrate into the insulator even if its energy is less than the barrier height. This phenomenon is referred to as tunneling. It is tunneling that explains how the Cooper pairs get through the insulator. The phase of the waves in each superconducting part of the junction may not necessarily be the same. The current flowing will depend on the relative phases of the waves in the two superconductors. The Josephson effect is the flow of a super current I_s given by

(a)

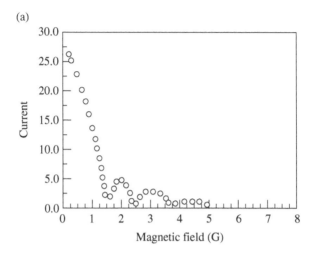

(b)

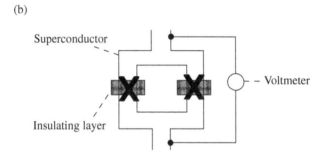

FIGURE 2.4 (a) Dependence of current through a Josephson junction on the strength of an applied dc magnetic field. (b) Schematic of a SQUID detector employing two junctions indicated by Xs. The magnetic field to be measured is applied perpendicular to the ring (Adapted from Ref. [1]).

$$I_s = I_c \sin \Delta\theta \qquad (2.8)$$

across the junction in the absence of an applied potential difference, where I_s is the maximum super current that the junction can support and $\Delta\theta$ is the phase difference between the wave functions of the Cooper pairs on each side of the junction.

These relative phases determine how the waves combine to give a net current. When the two waves are in phase, they combine to give the largest current. On the other hand, when the two waves are out of phase, that is, shifted by half a wave length with respect to each other, the combined wave gives no current. Phase shifts between these two extremes give an intermediate amount of current. It turns out that the relative phase of the waves in the two superconductors can be changed by the application of a magnetic field. Figure 2.4a shows how the current through the junction is changed as the strength of an applied magnetic perpendicular to the loop is increased [1].

The points where there is no current are the points where the waves from each super-conductor are a half wavelength out of phase. The current in a Josephson junction is very sensitive to a small magnetic field, and this is the basis of the SQUID magnetometer. Figure 2.4b shows a schematic illustration of a SQUID detector employing two junctions indicated by Xs in the figure.

2.3 FERROMAGNETISM

2.3.1 Theory of Ferromagnetism

Now, let us consider the question of why the individual atomic magnets align in some materials and not others. When a dc magnetic field is applied to a bar magnet, the magnetic moments tend to align with the direction of the applied field. In a crystal, each atom having a magnetic moment has a magnetic field about it. If the magnetic moment is large enough, the resulting large dc magnetic field can force a nearest neighbor to align in the same direction provided the interaction energy is larger than the thermal vibrational energy, kT, of the atoms in the lattice. The interaction between atomic magnetic moments is of two types, the so-called exchange interaction and the dipolar interaction. The exchange interaction is a purely quantum mechanical effect and is generally the stronger of the two interactions.

In the case of a small particle such as an electron, which has a magnetic moment, $g\mu_B$, the application of a dc magnetic field forces its spin vector to align such that it can have only two projections in the direction of the dc magnetic field, which are $\pm\frac{1}{2}g\mu_B$, where μ_B is the magnetic moment called the Bohr magneton and $g = 2.0023$ is the dimensionless gyromagnetic ratio of a free electron. The wave function representing the state $+\frac{1}{2}g\mu_B$ is designated α, and for $-\frac{1}{2}g\mu_{B\mu}$, it is β. The numbers $\pm\frac{1}{2}$ are called the spin quantum numbers m_s. For a two-electron system, it is not possible to specify which electron is in which state. The Pauli exclusion principle does not allow two electrons in the same energy level to have the same spin quantum numbers m_s. Quantum mechanics deals with this situation by requiring that the wave function of the electrons be antisymmetric, that is, it changes sign if the two electrons are interchanged. The form of the wave function that meets this condition is $(1/2)^{-1/2}[\Psi_A(1)\ \Psi_B(2) - \Psi_A(2)\ \Psi_B(1)]$. The electrostatic energy for this case is given by the expression z:

$$E = \int \left[\frac{1/2e^2}{r_{12}}\right]\left[\Psi_A(1)\Psi_B(2) - \Psi_A(2)\Psi_B(1)\right]^2 dV_1 dV_2 \qquad (2.9)$$

which involves carrying out an integration over the volume. Expanding the square of the wave functions gives two terms:

$$E = \int \left[\frac{e^2}{r_{12}}\right]\left[\Psi_A(1)\Psi_B(2)\right]^2 dV_1 dV_2 - \int \left[\frac{e^2}{r_{12}}\right]\left[\Psi_A(1)\Psi_B(1)\Psi_A(2)\Psi_B(2)\right]dV_1 dV_2$$

$$(2.10)$$

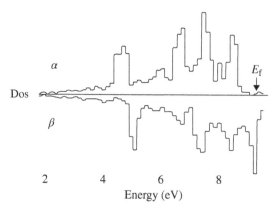

FIGURE 2.5   Calculated density of states at the Fermi level of nickel for the spin-up (α) and spin-down state (β) (Adapted from Ref. [2]).

The first term is the normal Coulomb interaction between the two charged particles. The second term, called the exchange interaction, represents the difference in the Coulomb energy between two electrons with spins that are parallel and antiparallel. It can be shown that under certain assumptions the exchange interaction between the spins S_1 and S_2 can be written in a much simpler form as $JS_1 \cdot S_2$, where J is called the exchange integral, or exchange interaction constant. This is the form used in the Heisenberg model of magnetism. For a ferromagnet, J is negative, and for an antiferromagnet, it is positive. The exchange interaction, because it involves the overlap of orbitals, is primarily a nearest neighbor interaction, and it is generally the dominant interaction. The other interaction, which can occur in a lattice of magnetic ions, called the dipole–dipole interaction, has the form

$$\frac{\mu_1 \cdot \mu_2}{r^3} - 3\frac{(\mu_1 \cdot r)(\mu_2 \cdot r)}{r^5} \qquad (2.11)$$

where r is a vector along the line separating the two magnetic moments μ_1 and μ_2 and r is the magnitude of this distance. The symbols in bold are vector quantities.

For ferromagnetism to exist in a semiconductor or metal, the density of states at the Fermi level must be different for the spin-down (β) states compared to the spin-up (α) states. As discussed in Chapter one, the Fermi level is the top-filled energy level in a solid, and the density of states is the number of energy levels in an interval of energy. A calculation of the density of states for the spin-up and spin-down state of nickel is shown in Figure 2.5 indicating a higher density of states at the Fermi level for the spin-down state [2].

The magnetization M of a bulk sample is defined as the total magnetic moment per unit volume. It is the vector sum of all the magnetic moments of the magnetic atoms in the bulk sample divided by the volume of the sample. It increases strongly at the Curie temperature, T_c, the temperature at which the sample becomes ferromagnetic, and it eventually becomes constant as the temperature is lowered further below

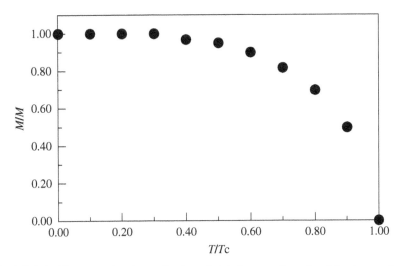

FIGURE 2.6 The saturation magnetization of nickel versus the ratio of the temperature to the Curie temperature (Adapted from Ref. [3]).

T_c. It has been found empirically that below the Curie temperature, the magnetization depends on the temperature as

$$M(T) = M(0)(1 - cT^{3/2}) \qquad (2.12)$$

where $M(0)$ is the magnetization at zero degrees Kelvin and c is a constant. This equation is known as the Bloch equation. Figure 2.6 illustrates a plot of the $M(T)/M(0)$ of Ni versus temperature [3].

Generally, for a bulk ferromagnetic material below the Curie temperature, the magnetization, M, is less than the magnetization the material would have if every individual atomic moment were aligned in the same direction. The reason for this is because of the existence of domains. Domains are regions in which all the atomic moments point in the same direction so that within each domain the magnetization is saturated, that is, it attains its maximum possible value. However, the magnetization vectors of different domains in the sample are not all parallel to each other. Thus, the sample has a total magnetization less than value for the complete alignment of all moments. Some examples of domain configurations are illustrated in Figure 2.7a. They exist because the magnetic energy of the sample is lowered by the formation of domains. Applying a dc magnetic field can increase the magnetization of a sample. This occurs by two processes. The first process occurs in weak applied fields when the volume of the domains that are oriented along the field direction increases. The second process dominates in stronger applied fields that force the magnetization of the domains to rotate toward the direction of the field. Both of these processes are illustrated in Figure 2.7b. Figure 2.8 shows a schematic plot of the magnetization curve of a ferromagnetic material. It is a plot of the total magnetization of the sample M versus the applied dc field strength, H. Initially, as H increases, M increases until a saturation point M_s is

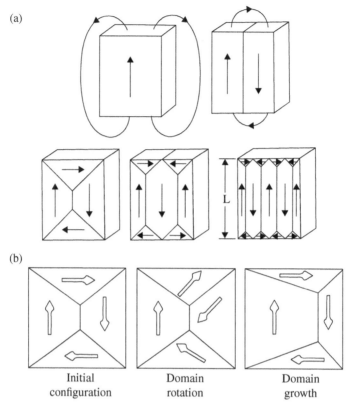

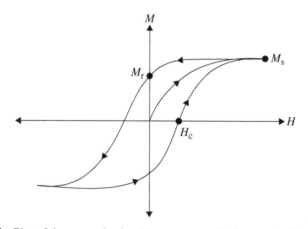

FIGURE 2.7 Illustration of some domain structures in ferromagnetic materials (a) and illustration of domain rotation and growth (b).

FIGURE 2.8 Plot of the magnetization M versus an applied magnetic field H for a hard ferromagnetic material, showing the hysteresis loop with the coercive field H_c, the remnant magnetization M_r, and the saturation magnetization M_s, indicated.

reached. When H is decreased from the saturation point, M does not decrease to zero magnetization; rather, it has a nonzero magnetization at zero magnetic field. This is called hysteresis. It occurs because the domains that were aligned with the increasing field do not return to their original orientation when the field is lowered. When the applied field H is returned to the zero, the magnet still has a magnetization referred to as the remnant magnetization, M_r. In order to remove the remnant magnetization, a field H_c has to be applied in the opposite direction to the initial applied field, as shown in Figure 2.8. This field, called the coercive field, causes the domains to rotate back to their original positions. The properties of the magnetization curve of a ferromagnetic have a strong bearing on the use of magnet materials, and there is much ongoing research to design permanent magnets having different kinds of magnetization curves. For permanent magnets, large hystereses are desirable, while magnets used in AC devices such as motors and generators need small or zero hysteresis.

In a single crystal, the magnetic interactions between the substituents of the lattice will be anisotropic and depend on the direction in the crystal. This means the magnetization will be different depending on the direction in which the dc magnetic field is applied. There is thus an energy in a ferromagnetic crystal that influences the magnetization depending on the direction in the lattice. This energy is called the anisotropy energy. For example, cobalt has a hexagonal close-packed structure. It consists of parallel planes of cobalt atoms having a hexagonal arrangement. The magnetization is much smaller for a magnetic field perpendicular to the close-packed plane (the easy direction) compared to the same field applied parallel to this plane. This is because the distance between the Co atoms is shorter in the plane than between the planes. Thus, the magnetic interactions are stronger between the atoms in the plane. The anisotropy energy for cobalt is given by

$$E(K) = K_1 \sin^2 \theta + K_2 \sin^4 \theta \qquad (2.13)$$

where θ is the angle between the magnetization and the easy axis. K_1 and K_2 are the anisotropy constants, which for cobalt at room temperature are 4.1×10^6 and $1 \times 10^6 \, \text{ergs/cm}^2$, respectively.

2.3.2 Magnetic Resonance

In Section 2.2.2, two methods of measuring magnetic properties of materials, AC susceptibility and SQUID magnetometry, were described. Here, we will discuss another method, magnetic resonance, that can be used to experimentally investigate magnetic materials.

When a magnetic field is applied to an atom or molecule having an unpaired electron, the energy of the $m_s = +1/2$ is greater than the $m_s = -1/2$, and the difference depends on the strength of the applied field H as $g\beta H$ where g and β are constants. In the electron spin resonance (ESR) method, radiation of a fixed frequency, ν, typically in the microwave region having energy, $h\nu$, is applied to the sample, and the magnetic field slowly increased until the separation between the spin states $g\beta H$ equals $h\nu$. When this occurs, there is a transition of spin-down electrons to the spin-up state, and microwave energy is absorbed. The situation is illustrated in Figure 2.2. The sample is in a microwave cavity, which concentrates the microwave radiation over the sample and is located between the poles of a magnet. For X band (9.2 GHz) and $g \cong 2$, the ESR absorption occurs for fields near 0.32 T. Superimposed on the sweep of the dc magnetic field is an ac modulation typically having a frequency of 100 kHz. This modulation is supplied by *rf* coils mounted on the side walls of the microwave cavity. The modulation of the absorption results in a time-varying output signal at the crystal diode, which changes phase by 180° at the peak of the absorption. Phase-sensitive detection is employed, which compares the phase of this output with that of a reference signal, enabling the derivative of the absorption to be detected. This reduces noise and enhances the sensitivity. A block diagram of a simple electron paramagnetic resonance spectrometer is presented in Figure 2.9. The klystron produces the microwaves that are transmitted via the waveguide to the microwave cavity. The cavity is between the poles of a dc electromagnet and contains the sample. When the sample absorbs, there is a reduction in reflected microwaves from the cavity, and this goes to the detector on the right side of the microwave bridge. When the sample in the microwave cavity is a magnetic material, the absorption of microwave is determined by the ferromagnetic properties and is called ferromagnetic resonance (FMR). FMR employs the same equipment as EPR.

However, in FMR, the *rf* energy is not absorbed by transitions of the electron spin from the spin-down state to the spin-up state but rather due to the precession of the total magnetization **M** of the sample about the applied dc magnetic field B_0. The energy is absorbed when the applied frequency is equal to the precessional frequency. This causes the direction of magnetization to flip. The equation of motion for a unit volume of the sample is

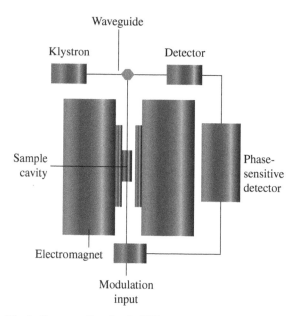

FIGURE 2.9 Block diagram of a simple ESR spectrometer. The sample is located in a microwave cavity between the poles of an electromagnet.

$$\frac{d\mathbf{M}}{dt} = \gamma (\mathbf{M} \times B_0) \tag{2.14}$$

where γ is the magnetogyric ratio. The resonance frequency of the FMR signal depends on the shape of the sample through the demagnetization factors N_x, N_y, and N_z, which relate the internal magnetic field to the applied field in three mutually perpendicular directions. For a flat rectangular ferromagnetic plate, the resonance condition for B_0 perpendicular to the plate is

$$\omega_0 = \gamma (B_0 - 4\pi \mathbf{M}) \tag{2.15}$$

For the magnetic field parallel to the surface of the plate, the resonance frequency is given by

$$\omega_0 = \gamma \left[B_0 (B_0 - 4\pi \mathbf{M}) \right]^{1/2} \tag{2.16}$$

It is seen that ω_0 depends on the orientation of the magnetic field with respect to the geometry of the sample. Different expressions are obtained for spheres and cylinders. The major characteristics that distinguish FMR signals from ESR absorptions are strong temperature dependence of ω_0 because of the dependence of $\mathbf{M}$ on temperature and temperature-dependent line widths, which generally increase with lowering temperature.

Another major difference is that FMR spectrum shows a signal centered at zero magnetic field. Figure 2.10 shows the spectrum of powders of copper-doped ZnO, which has been shown to be ferromagnetic at room temperature [4]. The spectrum

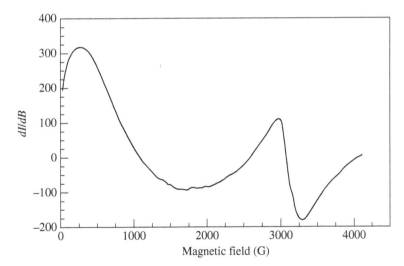

FIGURE 2.10 Ferromagnetic resonance signal obtained at room temperature in powders of ZnO doped with copper (Adapted from Ref. [4]).

consists of two lines, a broad line in the vicinity of 3000 G and a more intense line centered at zero magnetic field. The broad line at high field is due to the absorption of microwave radiation due to the reorientation of the magnetization vector. In a single crystal, the magnetic field position of the FMR signal depends on the orientation of the dc magnetic field with respect to important symmetry directions in the unit cell. The spectra shown here are from a collection of randomly oriented grains and are powder patterns representing the sum of spectra from all orientations of the dc magnetic field. The presence of the low-field nonresonant absorption signal is a well-established indication of ferromagnetism in materials. The signal occurs because the permeability in the ferromagnetic state depends on the applied magnetic field increasing at low fields to a maximum and then decreasing. Since the surface resistance depends on the square root of the permeability, the microwave absorption depends nonlinearly on the strength of the dc magnetic field resulting in a nonresonant derivative signal centered at zero field. This signal is not present in the paramagnetic state and emerges as the temperature is lowered to below the Curie temperature, T_c. It provides a unique signature of the presence of ferromagnetism.

2.4 ANTIFERROMAGNETISM

In the antiferromagnetic state, neighboring magnetic moments are oppositely aligned or antiparallel as shown in Figure 2.1c for a two-dimensional lattice. Antiferromagnetism occurs when the exchange interaction J in the Heisenberg representation is positive. The temperature dependence of the susceptibility is quite different from that in a ferromagnetic material. Manganese oxide (MnO) whose structure is shown in Figure 2.11 becomes

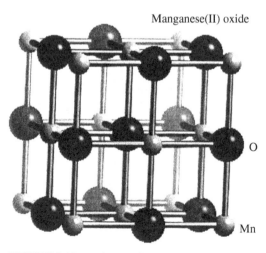

Manganese(II) oxide

O

Mn

FIGURE 2.11   Unit cell of manganese oxide, MnO.

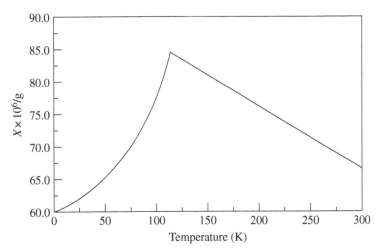

FIGURE 2.12 Measurement of the susceptibility versus temperature showing the antiferro-magnetic transition in MnO.

antiferromagnetic below 110K, which is referred to as the Neel temperature T_N. Figure 2.12 gives a plot of the temperature dependence of the susceptibility for powders of MnO showing a marked decrease in the susceptibility at 110K. In a later chapter, the effect of nanosizing on antiferromagnetic behavior will be discussed.

EXERCISES

2.1 Rhodium (Rh) has a face-centered cubic structure and is paramagnetic. The Rh atom has one unpaired electron, and the atom has a magnetic moment of 0.11 Bohr magnetons. Calculate the susceptibility of a 10mg sample of Rh in a 100G dc magnetic field at 300K.

2.2 Assuming the 10mg sample of Rh totally fills a coil whose frequency when empty is 150kHz, what would the frequency be when the sample is in the coil with no magnetic field applied?

2.3 If the electron spin resonance of a sample occurs at 3200G at X band, what field would it occur at K band?

2.4 The Curie temperature of Ni is 631K. The saturation magnetization at 0K is 510G. If a dc magnetic field of 3200G is applied perpendicular to a flat plate of Ni, what is the FMR frequency at 505 and 126K?

2.5 The Mn^{2+} ion in MnO gives a broad EPR signal at 3200G at room temperature. What would the signal be like below 110K. Explain your answer.

REFERENCES

1. F. J. Owens and C. P. Poole, in *The New Superconductors*, p. 68, Plenum Press, New York, 1996.
2. J. W. D. Connolly, *Phys. Rev.* 159, 415 (1967).
3. P. Weiss and R. Forrer, *Ann. Phys.* 5, 153 (1926).
4. F. J. Owens, *J. Mag. Mag. Mat.* 321, 3734 (2009).

3

PROPERTIES OF MAGNETIC NANOPARTICLES

3.1 SUPERPARAMAGNETISM

The existence of domains in magnetic materials as described in Chapter 2 leads to a hysteresis of the magnetization when it is measured as a function of increasing and decreasing applied DC magnetic field. When the size of an individual magnetic particle is reduced below some critical value, it is no longer energetically favorable to form domains and the particles have a single domain. For example, Fe_2O_3 nanoparticles become single domain below about 100 nm, and $CoFe_2O_4$ magnetic nanoparticles are single domain below 50 nm. The existence of single-domain nanoparticles results in the phenomenon of superparamagnetism where the magnetization as a function of increasing and decreasing magnetic field displays no hysteresis. In a single-domain ferromagnetic nanoparticle, the magnetic moment of the individual atoms is aligned in the same direction giving the particle a total magnetic moment, which is the sum of the individual moments of the constituent atoms. When a dc magnetic field is applied to a collection or powder of monodomain ferromagnetic nanoparticles, the total magnetic moment of each particle aligns with the dc magnetic field. As we will discuss in the following text, this alignment can be removed at some higher temperature where the thermal energy, kT, of the particles exceeds the energy of interaction, $\mu \cdot H$, between the magnetic moment of the particles and the applied dc magnetic field.

3.2 EFFECT OF PARTICLE SIZE ON MAGNETIZATION

For a single-domain nanoparticle, the total magnetic moment of the particle will be directly proportional to the number of magnetic atoms, N, in the particle, that is, $N\mu_i$ where μ_i is the magnetic moment of the individual atoms. For the case of an iron

Physics of Magnetic Nanostructures, First Edition. Frank J. Owens.
© 2015 John Wiley & Sons, Inc. Published 2015 by John Wiley & Sons, Inc.

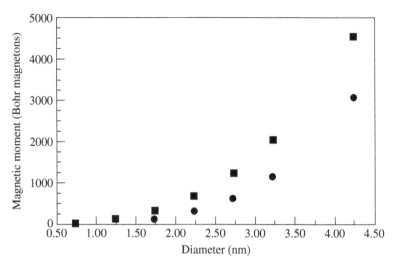

FIGURE 3.1 Calculated change of the total magnetic moment μ of an iron nanoparticle versus the diameter d of the particle assuming all the atoms of the particle are ferromagnetically ordered (■) and where surface atoms are not ordered (●).

nanoparticle, the increase in the total magnetic moment as a function of particle diameter can be calculated using the data in the table in Appendix A. The top curve in Figure 3.1 is a plot of the relative change of the total magnetic moment of an iron nanoparticle as a function of particle diameter in the monodomain region assuming that the magnetic moment of every iron atom in the particle is aligned in the same direction. However, this turns out not to be the case because the surface atoms of the particle, which have less nearest neighbors and therefore weaker magnetic interactions, become magnetically disordered. Since the percentage of atoms on the surface increases as the particle size decreases, this will contribute to a decrease in the magnetization with reduced size. The lower curve shows the dependence of the total magnetic moment on particle size assuming the magnetic moments of the atoms on the surface are not ferromagnetically ordered. When the particles become multidomain, the increase will begin to saturate at larger diameters. As discussed in Chapter 2, the interaction between the magnetic atoms of the particles that produces the net alignment of the atomic moments is the exchange interaction, $JS_1 \cdot S_2$, where the exchange integral, J, involves an overlap of the outer wave functions of nearest neighbor atoms. The strength of this interaction will determine the transition temperature. A stronger interaction means more thermal energy will be needed to cause the moments to become disordered and the ferromagnetism to disappear, resulting in a higher Curie temperature. Most ferromagnetic materials are metals. The lattice parameters of metals do not change much until the particle size is quite small, at which point it decreases slightly, meaning there should be a slight increase in the Curie temperature. However, if the magnetic metal nanoparticles have an oxide layer on them, as many do, the lattice parameters of the metal may change with particle size because of interfacial stress due to the mismatch of the lattice parameter of the metal and the metal oxide. For example, the lattice

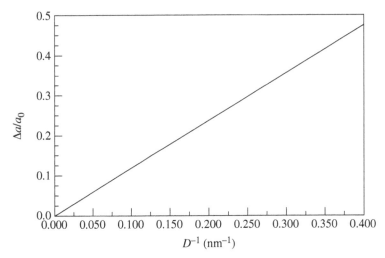

FIGURE 3.2 Plot of the increase of the lattice parameter of nickel nanoparticles versus the inverse of the particle size due to the interfacial stress caused by the mismatch of the lattice parameters of the Ni lattice and the NiO surface layer (Adapted from Ref. [1]).

parameter of nickel is 0.3524 nm and that of nickel oxide is 0.4203 nm. It has been shown by high-resolution transmission electron microscopy (HRTEM) that nickel nanoparticles have a 2–3 nm layer of nickel oxide on their surface [1]. The mismatch of the lattice parameters produces an interfacial stress that causes the lattice parameters of the nickel to expand at small particle sizes. Figure 3.2 shows a plot of the measured lattice parameter of nickel versus the reciprocal of particle size. The effect only occurs at very small particle sizes less than about 40 nm. This means there will be a reduction in the value of the exchange integral and therefore the magnetization with reduced particle size in a given dc magnetic field.

3.3 DYNAMICAL BEHAVIOR OF MAGNETIC NANOPARTICLES

Consider a powder of ferromagnetic nanoparticles cooled to a very low temperature, say, 4.2 K, in zero magnetic field. The direction of the total magnetic moment of each particle is frozen in a random arrangement, and the collection of particles will have a very low or zero net magnetization. If a dc magnetic field is applied and the susceptibility is measured as a function of increasing temperature, some fraction of the particles will begin to align with the field, and the susceptibility will increase with increasing temperature. If the particles are cooled to this low temperature in a dc magnetic field, the direction of maximum magnetization of the particles (often called the easy direction) will align with the applied dc field. This alignment will be frozen in at low temperatures. When the temperature is raised, a temperature is reached called the blocking temperature T_B, where the thermal energy, kT, of the particles is greater than the energy of interaction $\mu \cdot H$ of the moments of the particle, μ, with the dc magnetic field H. At this temperature, the direction of the magnetic moments of the particles begins to fluctuate about the direction of the dc magnetic field and the moments become disordered. The average time, τ, between reorientations is given by

$$\tau = \tau_0 \exp\left(\frac{\mu H}{k_B T}\right) \tag{3.1}$$

Thus, the blocking temperature T_B is a temperature below which the magnetic moments of the particles are locked in the direction of the magnetic field and above which they begin to randomly reorient. The blocking temperature can be measured by comparing the magnetization versus increasing temperature with zero field cooled (ZFC) and field cooled (FC) particles. Figure 3.3 illustrates a plot of the susceptibility of 8.5 nm particles of $CoFe_2O_4$ as a function of increasing temperature in FC and ZFC material in an applied magnetic field [2]. The temperature where the FC and ZFC susceptibilities diverge is the blocking temperature T_B. This temperature corresponds to the peak in the magnetization versus increasing temperature data for ZFC material. Figure 3.4 shows a plot of the magnetization versus increasing temperature for ZFC $CoFe_2O_3$ nanoparticles of different sizes measured in a 100 G applied field. Figure 3.5 shows a plot of the blocking temperature versus particle size for the $CoFe_2O_3$ nanoparticles. The blocking temperature increases with particle size because the total magnetic moment of the particles increases with size, making the interaction with the magnetic field stronger. Figure 3.6 shows how the saturation magnetization is affected as the particle size increases. The decrease in the magnetization with reduced particle size is a result of the decrease of the number of magnetic atoms in

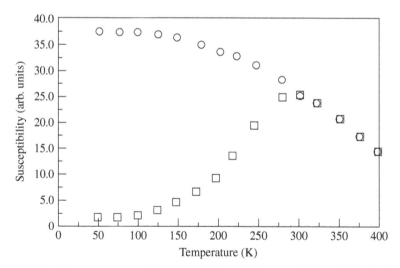

FIGURE 3.3 A measurement of the susceptibility of 8.5 nm particles of $CoFe_2O_4$ in field cooled (FC) (O) and zero field cooled (ZFC) (□) material as a function of increasing temperature (Adapted from Ref. [2]).

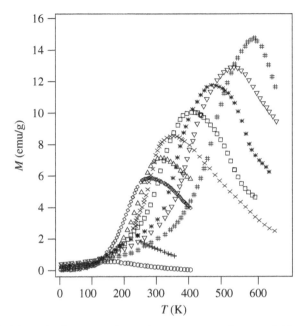

FIGURE 3.4 Magnetization versus increasing temperature for zero field cooled $CoFe_2O_4$ nanoparticles of different sizes measured in a 100 G field. The particle sizes range from 5 nm for the bottom curve to 30 nm for the top curve (Adapted from Ref. [2]).

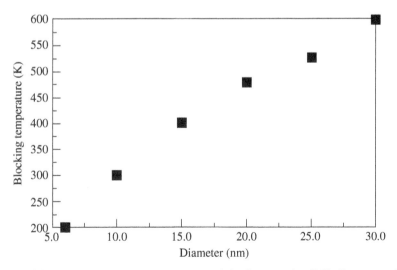

FIGURE 3.5 Blocking temperature versus particle diameter for $CoFe_2O_4$ nanoparticles obtained from the data in Figure 3.4 (Adapted from Ref. [2]).

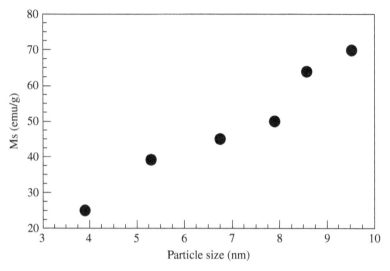

FIGURE 3.6   Saturation magnetization versus particle diameter for $CoFe_2O_4$ nanoparticles (Adapted from Ref. [2]).

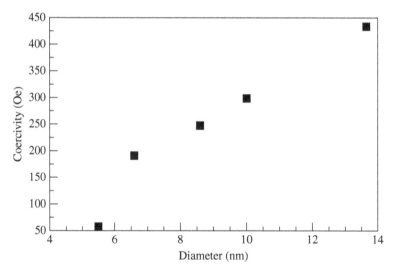

FIGURE 3.7   Coercivity versus diameter of the particles in the irreversible temperature region for nanoparticles of $MnFe_2O_4$ (Adapted from Ref. [2]).

the particle with reduced size and the increased number of atoms on the surface of the particles as discussed earlier. The coercivity in the irreversible temperature range also increases with particle size as shown in Figure 3.7. The measurements illustrate how particle size can be used to design the magnetic properties of nanoparticles.

3.4 MAGNETIC FIELD-ALIGNED PARTICLES IN FROZEN FLUIDS

As discussed earlier, the magnetic properties of powders of nanoparticles are dominated by the dynamical behavior of the particles that causes differences in the temperature dependence of the magnetization between magnetic FC samples and ZFC samples. It has been shown that magnetic nanoparticles can be aligned by suspending the particles in viscous liquids and cooling the liquid below the freezing point in a dc magnetic field [3]. The direction of the largest magnetization vector is aligned parallel to the direction of the cooling field. This alignment allows a study of the orientation dependence of the magnetic properties of the nanoparticles providing more information about the magnetic behavior of the particles. More importantly, it removes the effect of dynamical fluctuations on the magnetic properties, allowing a study of the intrinsic magnetic properties of the particles. For example, the orientation and temperature dependence of the ferromagnetic resonance (FMR) spectrum and the magnetization of iron nanoparticles suspended in the liquid crystal *N*-(*p*-methoxybenzylidene)-*p*-butylaniline (MBBA) and cooled below the freezing point (297 K) in a dc magnetic field have been studied [4]. This material was chosen because of its relatively high freezing temperature of 297 K, providing a wide range of temperatures to study the effects in the oriented nanoparticles. Figure 3.8 shows the FMR spectra recorded at 112 K for the dc magnetic field parallel to the direction of the cooling field for 75 nm iron particles frozen in MBBA and cooled in a 4000 G magnetic field. The in-field solidification locks the direction of maximum magnetization of the particles parallel to the direction of the applied dc magnetic field. The spectrum consists of two lines, a very low-field derivative signal centered at zero magnetic field and a higher-field FMR signal. The presence of the low-field

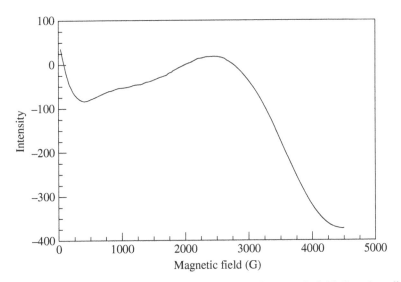

FIGURE 3.8 Ferromagnetic resonance signal at 112 K for dc magnetic field aligned parallel to the direction of the cooling field for iron nanoparticles in frozen MBBA (Adapted from Ref. [4]).

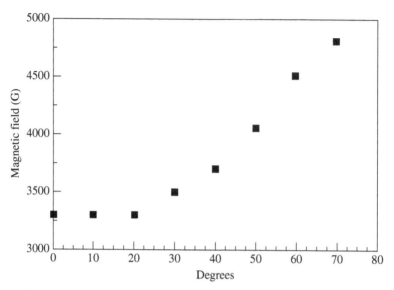

FIGURE 3.9 Dependence of the magnetic field position of the FMR signal of iron nanoparticles on the orientation of the dc magnetic field with respect to the direction of the cooling field at 112 K (Adapted from Ref. [4]).

nonresonant absorption signal is a well-established indication of ferromagnetism in materials as has previously been discussed in Chapter 2. Figure 3.9 shows a plot of the dependence of the magnetic field position of the higher magnetic field FMR signal on the orientation of the dc magnetic field with respect to the direction of the cooling field at 112 K. For a particle having uniaxial symmetry, the dependence of the field position of the FMR signal on the angle is given by [5]

$$H_r = H_0 - H_A \left(\frac{1}{2}\right)(3\cos^2\theta - 1) \tag{3.2}$$

where the anisotropy field is $H_A = 4|K|/M$, K is the anisotropy constant, and M is the magnetization. The angle, θ, is between the direction of maximum magnetization and the applied dc field, H_r is the magnetic field at the center of the FMR signal, and H_0 determines the g value. By fitting the data in Figure 3.9 to Equation 3.2, H_0 and H_A are determined to be 4210 and 910 G, respectively, at 112 K. The intensity of the low-field nonresonant signal and the FMR signal decreases as the dc magnetic field is rotated away from the direction of the cooling field. The intensity, field position, and line width of the FMR signal below the freezing point are independent of temperature, which means below 297 K, K, M, and H_0 are independent of temperature. The sample magnetization depends on the orientation of the dc magnetic field in frozen state with respect to the direction of the cooling field. Figure 3.10 shows a plot of the relative magnetization at 153 K normalized to its value at a 3 kG magnetic field for the field parallel and perpendicular to the direction of the cooling field.

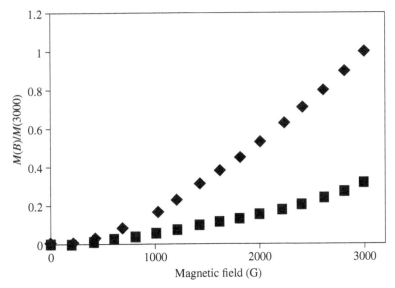

FIGURE 3.10 Plot of the relative magnetization normalized to its value at 3000 G for the dc magnetic field parallel (♦) and perpendicular (■) to the direction of the cooling magnetic field for iron nanoparticles in frozen MBBA (Adapted from Ref. [4]).

This result is consistent with the decrease in the intensity of the low-field signal and the FMR signal as the field is rotated away from the direction of the cooling field, indicating that the intensity of the FMR signal is proportional to the magnetization of the sample.

Studies of field-aligned Ni nanoparticles in frozen fluids show anomalous behavior compared to iron nanoparticles. Figure 3.11 illustrates a representative scanning electron microscope (SEM) image of the particles [6]. The insert in the figure shows the distribution of particle sizes obtained from the SEM image. The average particle size obtained from the SEM data is 109 nm. Figure 3.12 shows the FMR spectrum recorded at 104 K for the magnetic field parallel to the direction of the cooling field. The spectrum consists of two lines, a very low-field derivative signal centered at zero magnetic field and a higher-field FMR signal as observed in the case of iron discussed previously. Similar to the iron nanoparticles, the magnetic field at which the resonance occurs depends on the orientation of the dc magnetic field with respect to the direction of the cooling field.

Figure 3.13 shows that the FMR spectrum displays an unusual temperature dependence. The figure shows the spectra at 275 and 115 K for the magnetic field parallel to the cooling field, showing a very marked decrease in the intensity of the spectrum at the lower temperature. Figure 3.14 illustrates a plot of the temperature dependence of the intensity of the low-field derivative signal for the parallel orientation of the particles cooled below the freezing point of MBBA in a 0.4 T dc magnetic field. A similar temperature dependence is also observed for the high-field FMR signal. These data seem to suggest that the magnetization is decreasing

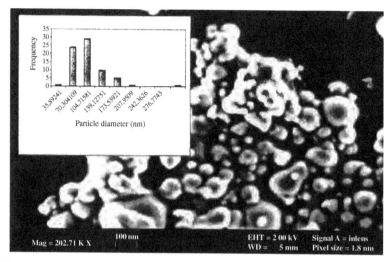

FIGURE 3.11 Scanning electron microscope picture of Ni nanoparticles. The insert shows the distribution of particle sizes obtained from the image (Adapted from Ref. [6]).

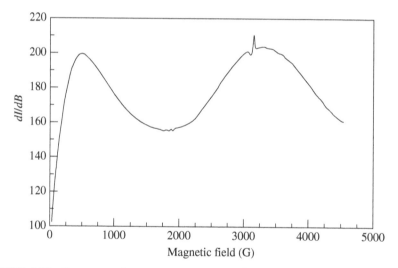

FIGURE 3.12 Ferromagnetic resonance spectrum of Ni nanoparticles suspended in the liquid crystal MBBA and cooled below the freezing point of the liquid in a 4 KG magnetic field. The signal is recorded for the dc magnetic field parallel to the direction of the cooling field at 104 K (Adapted from Ref. [6]).

as the temperature is lowered. However, as we will discuss in the following text, this is likely an artifact resulting from the relative size of the microwave skin depth of the Ni nanoparticle and particle size. Figure 3.15 shows a measurement of the temperature dependence of the surface resistance of the Ni particles below room

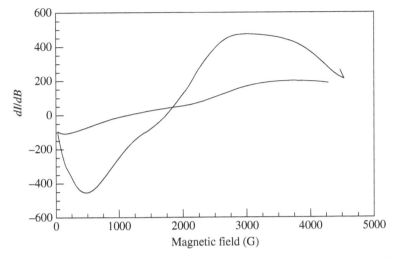

FIGURE 3.13 Ferromagnetic resonance spectra for H parallel to the cooling field of Ni in frozen MBBA at 275 and 115 K (smaller spectrum) showing a marked reduction in intensity as temperature is lowered (Adapted from Ref. [6]).

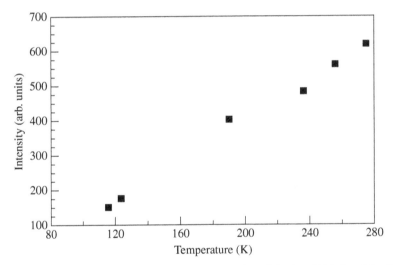

FIGURE 3.14 Temperature dependence of the intensity of the low-field signal for the dc magnetic field parallel to the direction of the cooling field of Ni in frozen MBBA (Adapted from Ref. [6]).

temperature in the liquid crystal for the magnetic field parallel to the direction of the cooling field. The same measurement on the liquid crystal without the particles shows no decrease, indicating that the decrease is associated with the particles.

An examination of the dependence of the FMR spectra on the angle of the DC field with respect to the direction of the cooling field at a number of temperatures

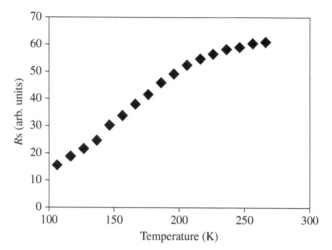

FIGURE 3.15   Temperature dependence of the surface resistance of the oriented Ni nanoparticles in the frozen fluid (Adapted from Ref. [6]).

below the freezing point of the liquid shows that there are no changes in the orientation. The particles at all temperatures below the freezing points of the liquid crystal remain aligned such that the direction of maximum magnetization is in the direction of the cooling field. This means that the observed increase in the FMR signal intensity between 100 and 300 K is intrinsic to the nanoparticles and not associated with dynamical fluctuations of the particles. Figure 3.15, which shows a decrease in the surface resistance with lowering temperature, suggests a possible explanation for the reduction in intensity of the FMR signal with lowering temperature. The surface resistance is proportional to $1/(\sigma)^{1/2}$ where σ is the conductivity. As the temperature is lowered, σ increases and the surface resistance decreases. The skin depth, which measures the depth of penetration of the microwaves into the material, is also proportional to $1/(\sigma)^{1/2}$ having the specific form

$$\delta = \frac{1}{\left[\pi U \sigma f\right]^{1/2}} \tag{3.3}$$

where U is the permeability, which in the ferromagnetic state depends on the strength of applied magnetic field, and f is the frequency in cycles per second. The conductivity increases with lowering temperature, and thus, the penetration depth of the microwaves into the sample decreases, resulting in a decrease in the FMR signal intensity. However, this will only occur if the skin depth is smaller than the size of the particle. Taking U in nickel at 3000 G as 10^{-5} H/m and σ of nickel as 1.2×10^{7} mhos/m, the skin depth at 9.2×10^{9} CPS is estimated to be 5 nm, which from the particle size distribution in Figure 3.11 is significantly smaller than the size of most of the particles. If the particle size is much smaller than the skin depth, then no increase in the FMR signal strength with increasing temperature will be observed, and intrinsic magnetic effects or dynamical effects can be measured by FMR. On the other hand, if the average particle size is greater than the skin depth, as is the case here, the intensity of the FMR signal will be temperature dependent. This underlines an important point that if FMR measurements are to be used to study intrinsic magnetic behavior of magnetic nanoparticles, particularly the temperature dependence, the particle size must be smaller than the skin depth of the microwave probing signal.

3.5 MAGNETISM INDUCED BY NANOSIZING

The rhodium (Rh) atom has 45 electrons circulating about the nucleus in the [Kr] $5S^14d^8$ electron configuration, so it has an unpaired electron and thus a magnetic moment. The Kr refers to the closed shell configuration of krypton. Macroscopic rhodium has a face-centered cubic structure similar to the MnO structure shown in Chapter 2. The material is paramagnetic at all temperatures showing no indication of ferromagnetism. Theoretical calculations of Rh nanoparticles having 13 atoms predicted that the nanoparticle would have a magnetic moment of $21\mu_B$ where μ_B is the Bohr magneton. The calculations indicated the possibility of an icosahedral structure that is different from the macroscopic face-centered cubic structure. It was subsequently experimentally shown that Rh nanoparticles have a net magnetic moment, and the magnitude of the moment depends on the number of atoms in the particle [7]. Figure 3.16 shows a plot of some of the experimental measured magnetic moments of the particles versus the number of atoms in the particle. The reason for the onset of ferromagnetism in the nanoparticles may be a result of the change of crystal structure, where the new structure has smaller separations between the atoms and therefore larger

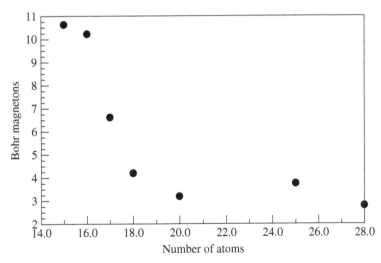

FIGURE 3.16 Plot of the magnetic moment of rhodium nanoparticles versus the number of atoms in the particle (Adapted from Ref. [7]).

exchange interactions. This result provides a unique demonstration of how nanosizing can alter properties making a nonmagnetic material magnetic.

3.6 ANTIFERROMAGNETIC NANOPARTICLES

In the antiferromagnetic state, the magnetic moments of the individual atoms of the material have an antiparallel arrangement as shown in Figure 2.1c for a two-dimensional system. In the antiferromagnetic state, the material has no net magnetic moment and the susceptibility will be zero. Antiferromagnetic nanoparticles display unusual behavior compared to ferromagnetic nanoparticles. In 1961, Néel proposed that antiferromagnetic nanoparticles could display weak ferromagnetism and superparamagnetism. It was seen earlier that as the particle size decreases, the percentage of atoms on the surface of the particle increases. Néel proposed that for antiferromagnetic nanoparticles, the antiparallel orientation of the moments on the surface would be lost and that the moments of the surface atoms would actually align ferromagnetically parallel to the axis of the antiferromagnetic alignment in the interior of the particle. This idea means that an antiferromagnetic nanoparticle can be viewed as having an inner antiferromagnetic core surrounded by an outer shell of ferromagnetically ordered spins. Experimental measurements in nickel oxide (NiO) confirm this model. Bulk NiO has a rhombohedral crystal structure and becomes antiferromagnetic below 523 K. Above that temperature, it is paramagnetic and the crystal structure is cubic. Figure 3.17 shows a measurement of the dc susceptibility of NiO nanoparticles versus temperature for particles of two different sizes, which

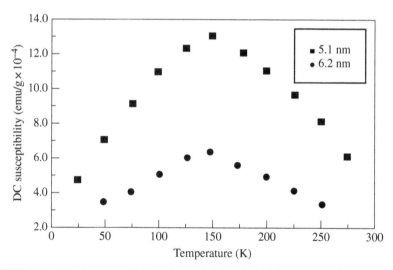

FIGURE 3.17 The dc susceptibility of zero field cooled NiO nanoparticle of two different sizes measured in a 100 G field as a function of increasing temperature (Reprinted with permission from Ref. [8]).

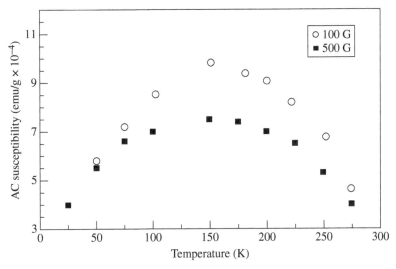

FIGURE 3.18 A measurement of the temperature dependence of the ac susceptibility for a NiO nanoparticle in two different magnetic fields (Reprinted with permission from Ref. [8]).

were ZFC, and measured in a field of 100 G [8]. The presence of a nonzero suscepti-bility means the antiferromagnetic particles are displaying ferromagnetic behavior. For particles greater than 100 nm, the susceptibility is zero. For the smaller particles, the figure shows that the susceptibility increases with increasing temperature to a maximum and then decreases. The dc magnetic field at which the maximum occurs shifts to higher temperature as the particle size gets smaller, and the magnitude of the susceptibility increases with reduced particle size. The data show clearly that the behavior is not a manifestation of a blocking temperature that increases as the par-ticle size gets larger, as shown in Figure 3.5, contrary to the behavior shown in Figure 3.17. The increase in the magnitude of the susceptibility is a result of the increase of the number of Ni atoms on the surface having ferromagnetically ordered spins. Figure 3.18 shows a measurement of the ac susceptibility for a particle of one size in two different dc magnetic fields. It is seen that the peak shifts slightly to lower temperature with increasing dc magnetic field. The relationship of the peak temper-ature to the magnetic field has been shown to follow an equation of the form [9]

$$B = c\left[1 - \frac{T}{T_s}\right]^{3/2} \tag{3.4}$$

This is referred to as the de Almeida–Thouless line, and it describes the relation of the field and temperature for a transition to a spin glass state, which is essentially a state where the magnetic moments of the surface atoms are disordered. Thus, the peak temperature is the temperature at which the surface moments begin to lose their ferromagnetic order.

3.7 MAGNETORESISTIVE MATERIALS

Magnetoresistance refers to a phenomenon where a dc magnetic field changes the resistance of a material. The materials have a number of application possibilities, such as in magnetic recording heads or as sensing elements in magnetometers. The perovskite-like material $LaMnO_3$ has manganese in the Mn^{3+} valence state. If the La^{3+} is partially replaced with ions having a valence of 2+, such as Ca, Ba, Sr, Pb, or Cd, some Mn^{3+} ions transform to Mn^{4+} to preserve the electrical neutrality. This mixed-valence system has been shown to exhibit very large magnetoresistive effects. The unit cell of the crystal is sketched in Figure 3.19. The particular system, $La_{0.67}Ca_{0.33}MnO_x$, displays more than a thousandfold change in resistance with the application of a 6 T dc magnetic field [10]. It is referred to as a colossal magneto-resistive (CMR) material. The materials generally undergo a paramagnetic to fer-romagnetic transition at some temperature. The transition temperature depends on the kind of dopant and its concentration. When the grain sizes of these materials are nanometers, there are dramatic changes in the magnetoresistance. Figure 3.20 shows a plot of the log of the magnetoresistance of $La_{0.9}Te_{0.1}MnO_3$ at 100 K in a 0.5 T dc magnetic field as a function of the grain size [11]. The magnetoresistance MR is defined as

$$MR(\%) = \frac{\Delta\rho}{\rho(0)} = \left(\frac{\rho(0) - \rho(H)}{\rho(0)}\right) \times 100\% \qquad (3.5)$$

where $\rho(0)$ is the resistivity when no magnetic field is applied and $\rho(H)$ is the resis-tivity when a field H is applied. Figure 3.20 shows that when the grain size of the CMR is below 50 nm, there is a very pronounced increase in the magnetoresistance.

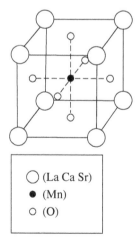

FIGURE 3.19 Crystal structure of $LaMnO_3$ that displays colossal magnetoresistance when the La site is doped with Ca or Sr.

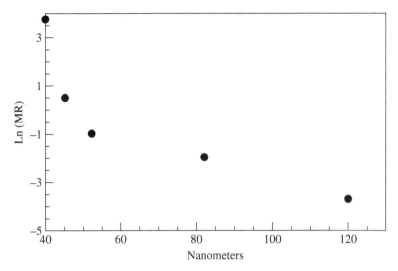

FIGURE 3.20   Plot of the logarithm of the magnetoresistance of $La_{0.9}Te_{0.1}MnO_3$ at 100 K in a 0.5 T dc magnetic field as a function of the particle diameter (Adapted from Ref. [10]).

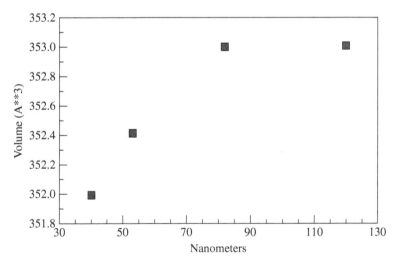

FIGURE 3.21   Plot of the volume of the unit cell of $La_{0.9}Te_{0.1}MnO_3$ versus particle diameter.

Similar results have been obtained in other CMR materials such as $La_{2-2x}Ca_{1+2x}Mn_2O_7$. It was also observed that the resistivity in zero field, $\rho(0)$, increased with reduced grain size. The Curie temperature and the magnetization at a given temperature are also observed to increase as the grain size decreases. A possible explanation of this is given in Figure 3.21, which is a plot of the volume of the unit cell measured by x-ray diffraction versus the grain size. This indicates that the Mn–Mn separation decreases,

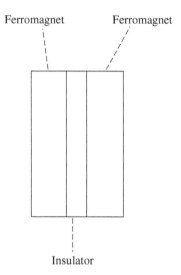

Ferromagnet   Ferromagnet

Insulator

FIGURE 3.22 Illustration of a magnetic tunnel junction consisting of two ferromagnetic materials separated by a thin insulating barrier layer.

and therefore, the interatomic exchange integral J increases, resulting in an increase in the magnetization and the Curie temperature.

Tunneling between the particles is involved in the mechanism of conductivity. The large increase in the magnetoresistance below 50 nm may be due to the formation of nanosized magnetic tunnel junctions formed by the contact of the grains of the CMR material. Magnetic tunnel junctions consist of two ferromagnetic metal electrodes separated by a small insulating barrier as shown in Figure 3.22. The application of a small voltage between the electrodes causes electrons to tunnel from the negative electrode to the positive electrode. When small magnetic fields are applied parallel to the electrodes, the tunneling conductance in the junction depends on the relative orientation of the magnetization in the two electrodes. If the magnetization vectors are parallel, the conductance is a maximum or the resistance is a minimum. When the magnetization vectors are antiparallel, the resistance is maximum. A more detailed explanation of tunneling will be presented in Chapter 4. The tunneling magnetoresistance ratio (TMR) is defined as

$$\text{TMR} = \frac{R_{ap} - R_p}{R_p} \tag{3.6}$$

where R_{ap} is the junction resistance when the magnetizations are antiparallel and R_p the resistance when they are parallel. Tunnel junctions have been fabricated from CMR materials where both electrodes are $La_{0.66}Sr0_{.3}MnO_3$. With a small voltage of 1.0 mV across the electrodes, a TMR ratio of 1800% was observed at 4.2 K. It was also observed that the TMR ratio increased as the size of the junction was decreased.

The qualitative similarity of this behavior to that observed in collections of particles of $La_{0.9}Te_{0.1}MnO_3$ when it is nanosized suggests the possibility of the formation of tunnel junctions between the particles.

EXERCISES

3.1 A spherical iron nanoparticle having a diameter of 4.83 nm contains 2869 iron atoms. 28.3% of these are on the surface. Taking the magnetic moment of iron as 2.22 Bohr magnetons, calculate the total magnetic moment of the particle.

3.2 Calculate the total magnetization of the above particle.

3.3 If a powder of the above particles is cooled to 0 K in the absence of a magnetic field and then a 200 G magnetic field is applied and the material warmed to room temperature, describe how the magnetization of the material behaves with increasing temperature.

3.4 Estimate the blocking temperature of the particles in Problem 3.3.

3.5 Ten milligrams of the above particles are suspended in an organic liquid that freezes at −40°C. The liquid is then cooled below the freezing point in a 500 G dc magnetic field. What is the total magnetic moment of the liquid below the freezing point? What is the total magnetic moment of the liquid if it is cooled in zero magnetic field?

REFERENCES

1. R. Relinghaus et al. *Eur. Phys. J.* 16, 249 (2001).
2. C. R. Vestal and Z. J. Zhang, *Int. J. Nanotechnol.* 1, 240 (2004).
3. F. J. Owens, *J. Phys. Chem.* 64, 2289 (2003).
4. F. J. Owens, *Nanosci. Technol. Lett.* 3, 1 (2011).
5. R. S. de Biasi and T. C. Devezas, *J. Appl. Phys.* 49, 2466 (1978).
6. F. J. Owens and V. Stepanov, *J. Exp. Nanosci.* 3, 141 (2008).
7. A. J. Cox et al. *Phys. Rev.* B49, 12295 (1994).
8. S. D. Tiwari and K. P. Rajeev, *Phys. Rev.* B72, 104433 (2005).
9. J. R. L. de Almeida and D. J. Thoules *J. Phys.* A11, 983 (1978).
10. R. Mahendrian et al. *App. Phys. Lett.* 66, 223 (1995).
11. Y. Yang et al. *Solid State Commun.* 131, 393 (2004).

4

BULK NANOSTRUCTURED MAGNETIC MATERIALS

4.1 FERROMAGNETIC SOLIDS WITH NANOSIZED GRAINS

The diverse applications of magnets require the magnetization curve to have different properties. Magnets used in transformers and rotating electrical machinery are subjected to rapidly ac magnetic fields, so they repeat their magnetization curves many times a second, causing a loss of efficiency and a rise in the temperature of the magnet. The rise in temperature is due to frictional heating from domains as they continuously vary their orientations. The amount of loss during each cycle, meaning the amount of heat energy generated during each cycle around a hysteresis loop, is proportional to the area enclosed by the loop. In these applications, small or zero coercive fields are required to minimize the enclosed area. Such magnets are called soft magnetic materials. On the other hand, in the case of permanent magnets used as a part of high-field systems, large coercive fields are required, and the widest possible hysteresis loop is desirable. Such magnets are called hard magnets. High saturation magnetizations are also needed in permanent magnets.

Nanostructuring of bulk magnetic materials can be used to design the magnetization curve. Amorphous alloy ribbons having the composition $Fe_{73.5}Cu_1Nb_3Si_{13.5}B_9$ prepared by a roller method, and subjected to annealing at 673–923 K for 1 h in inert gas atmospheres, were composed of 10 nm iron grains in solid solutions. Such alloys have a saturation magnetization M_s of 1.24 T, a remnant magnetization M_r of 0.67 T, and a very small coercive field H_c of 0.53 A/m. Nanoamorphous alloy powders of $Fe_{69}Ni_9CO_2$ having grain sizes of 10–15 nm prepared by decomposition of solutions of $Fe(CO)_5$, $Ni(CO)_4$, and $Co(NO)(CoO)_3$ in the hydrocarbon solvent decalin ($C_{10}H_{18}$) under an inert gas atmosphere showed almost no hysteresis in the magnetization curve. Figure 4.1 presents the magnetization curve for

Physics of Magnetic Nanostructures, First Edition. Frank J. Owens.
© 2015 John Wiley & Sons, Inc. Published 2015 by John Wiley & Sons, Inc.

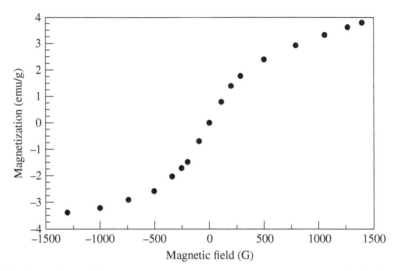

FIGURE 4.1   Reversible magnetization curve for nanosized powders of a Ni–Fe–Co alloy that exhibits no hysteresis. An oersted corresponds to 10^{-4}T (Reproduced with permission from Ref. [1]. Copyright 2000, Cambridge University Press).

this material [1]. As discussed earlier, a magnetic material that has no hysteresis at any temperature is said to be superparamagnetic.

The strongest known permanent magnets are made of neodymium, iron, and boron. They can have remnant magnetizations as high as 1.3 T and coercive fields as high as 1.2 T. The effect of the size of the nanoparticle grain structure on $Nd_2Fe_{14}B$ has been investigated. The results, shown in Figures 4.2 and 4.3, indicate that in this material the coercive field decreases significantly below about 40 nm and the remnant magnetization increases [2]. Another approach to improving the magnetization curves of this material has been to make nanoscale compositions of hard $Nd_2Fe_{14}B$ and the soft α–phase of iron. Measurements of the effect of the presence of the soft iron phase mixed in the hard material confirm that the remnant field can be increased by this approach. This is believed to be due to the exchange coupling between the hard and soft nanoparticles, which forces the magnetization vector of the soft phase to be rotated to the direction of the magnetization of the hard phase.

The size of magnetic nanoparticles in the composite affects the value of M_s at which the magnetization saturates. Figure 4.4 shows the effect of particle size on the saturation magnetization of zinc ferrite, illustrating how the magnetization increases significantly below a grain size of 20 nm [3]. Thus, decreasing the particle size of a composite material can increase the magnetization. It should be noted that this increase in magnetization with reduced grain size of the bulk is opposite to what is observed for powders of magnetic nanoparticles. Perhaps this is a result of the dense packing of the grains in composite materials that does not allow the moments of the surface atoms to become disordered because of interparticle interactions.

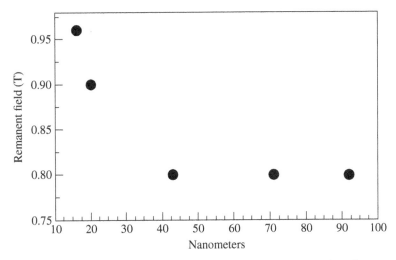

FIGURE 4.2 Dependence of the remnant magnetization M_r on the particle diameter, d, of the grains that form the structure of a Nd–B–Fe permanent magnet (Adapted from Ref. [2]).

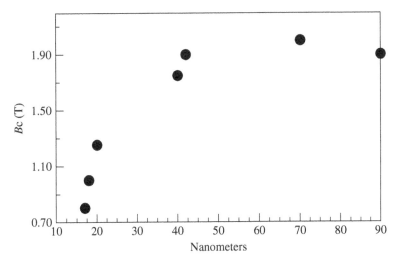

FIGURE 4.3 Dependence of the coercive field B_c (i.e., H_c) on the granular particle diameter, d, of a Nd–B–Fe permanent magnet (Adapted from Ref. [2]).

4.2   LOW-DIMENSIONAL MAGNETIC NANOSTRUCTURES

4.2.1   Magnetic Quantum Wells

If one dimension is reduced to the nanometer range and the other two remain large, the structure is called a quantum well. The most obvious example of such a structure is a thin film that has nanometer thickness. The properties of thin ferromagnetic films

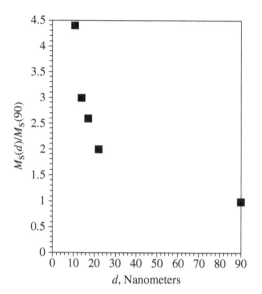

FIGURE 4.4 Dependence of the saturation magnetization M_s of zinc ferrite on the granular particle size, d, normalized to the value $M_s(90)$ for a 90 nm grain. (From Ref. [3]. © IOP Publishing. Reproduced by permission of IOP Publishing. All rights reserved).

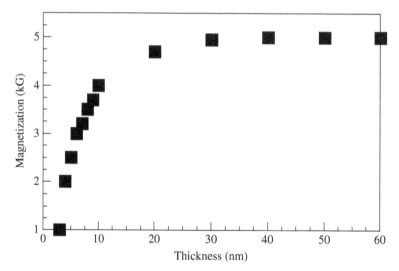

FIGURE 4.5 Dependence of the saturation magnetization on the thickness of Ni films that were electrochemically deposited on a copper substrate (Adapted from Ref. [4]).

have been studied for many years long before the word nanotechnology came into use. Figure 4.5 shows the dependence of the saturation magnetization on the thickness of an electrodeposited nickel film on a copper substrate [4]. Below about 30 nm, there is the onset of a substantial reduction in the saturation magnetization. The

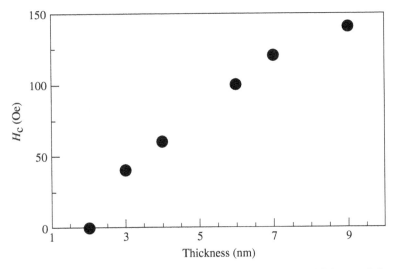

FIGURE 4.6 Dependence of the coercivity of the Ni film on the thickness of the films (Adapted from Ref. [4]).

reason for this decrease is because of the increased fraction of magnetic atoms on the surface below 40 nm that lose their ferromagnetic order. Figure 4.6 shows the effect of film thickness on the coercivity [4]. Notice that at a thickness of 2 nm, there is no hysteresis, meaning the film is superparamagnetic. Multilayered films of ferromagnetic materials have been developed into a number of devices such as spin valves that will be discussed later.

Ferromagnetic resonance (FMR) has also been used to study ferromagnetic wells. The studies have been made on ultrathin films of Au/Fe deposited by molecular beam epitaxy on the (001) surface of thin crystals of gallium arsenide. A description of molecular beam epitaxy is provided in Chapter 9. In this method, a vapor of Fe is deposited on a substrate followed by deposition of a vapor of Au. The process is carried out in a vacuum chamber, and the thickness of the layers is monitored by electron diffraction. The films were 20 atomic layers thick of gold on top of a 10 atomic layers of iron. Figure 4.7 shows the FMR spectra at a number of different temperatures for the dc magnetic field in the plane of the film parallel to the [110] direction of the GaAs substrate [5]. The spectra are asymmetric and shift to lower magnetic fields as the temperature is lowered. The line width also increases with decreasing temperature. As discussed earlier, these effects are characteristic of FMR spectra. When the measurements were made with the dc magnetic field in the plane of the film and perpendicular to [110], the resonance shifted to higher magnetic field values. The temperature dependence of the lines is due to the temperature dependence of K/M. The dependence of the line position for the case of axial symmetry, given by Equation 3.2, shows that when H is parallel to the direction of maximum magnetization, the line shifts to lower magnetic field. When it is perpendicular to that direction, the line would shift up with temperature. A measurement of

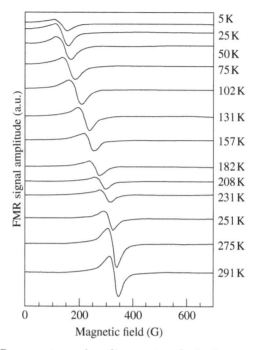

FIGURE 4.7   FMR spectra at a number of temperatures for the dc magnetic field in the plane of an Au/Fe film (Reproduced with permission from Ref. [5]. Copyright © 2007, Springer-Verlag Berlin Heidelberg).

the temperature dependence of the magnetic field at the center of the FMR resonance allows a measurement of the temperature dependence of K and M.

Layered transition metal dichalcogenides have the potential to enable fabrication of magnetic quantum wells. A typical member of this family of materials is vanadium disulfide (VS_2) that in its bulk form is ferromagnetic. The V^{4+} ion has a large magnetic moment of $3\mu_B$. One crystal structure of this material consists of parallel layers of VS_2 where the S has a charge of -2. The layers are weakly bonded to each other in the lattice by van der Waals forces. There are two stable phases for this material designated as the $1T$ and $2H$ phases that differ by the local arrangement of the sulfur atoms about the metal ion. In the $2H$ phase, the sulfur atoms form a trigonal prism array around the metal. In the $1T$, the atoms form an octahedral arrangement about the metal ion. Because of the weak interaction between the VS_2 layers, it has been possible to produce thin nanosheets of VS_2 by exfoliation [6]. Exfoliation is a method that increases the separation between the layers by incorporating other materials between them such as NH_3 in this case. When sonication is applied to a liquid containing this material, nanosheets having less than 5 VS_2 layers can be separated out. While bulk VS_2 is a ferromagnetic metal, evidence for ferromagnetism in the nanosheets remains to be determined. However, there are theoretical predictions that such sheets should be ferromagnetic [7].

4.2.2 Magnetic Quantum Wires

When two dimensions are reduced to nanometer length and one remains large, it is a quantum wire. Single-walled carbon nanotubes, discussed in the next chapter, are an example of such a wire. Magnetic wires made of Fe_3O_4 have been fabricated.

Changing the diameter can be used to control the magnetic properties of the wires. Figure 4.8 shows a plot of the coercivity of the wires as a function of the diameter [8]. Below about 50 nm, the coercivity is zero, meaning that the magnetization versus magnetic field displays no hysteresis and the wires are superparamagnetic. Powders of magnetic wires display dynamical fluctuations. Figure 4.9 shows a plot of the blocking temperature of the hematite wires versus an applied dc magnetic field [8]. This plot is determined from a measurement of the magnetization versus dc magnetic field for zero field and field cooled wires using different magnetic fields as discussed in Chapter 3 for the magnetic nanoparticles.

Copper oxide (CuO) undergoes a transition from a paramagnetic state to an antiferromagnetic state at 230 K. When CuO is fabricated into nanorods having diameters of 30–40 nm and lengths of 100–200 nm, it displays ferromagnetism. Figure 4.10 shows a plot of the measured magnetization versus dc magnetic field at room temperature showing a substantial magnetization [9]. This is another example of how nanosizing can dramatically alter the magnetic properties of a material making a nonferromagnetic material ferromagnetic. It was seen earlier that antiferromagnetic nanoparticles such as NiO become ferromagnetic at small sizes because of the increased number of atoms on the surface that tend to order ferromagnetically. The same effect is occurring in the CuO nanorods. The magnetization is stronger than an equivalent-sized nanoparticle because the rods have a greater surface area.

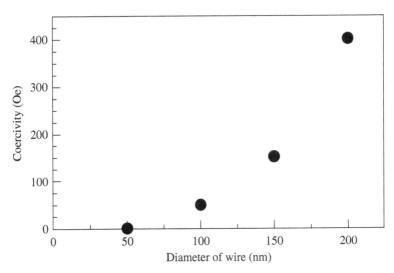

FIGURE 4.8 Dependence of the coercivity of an Fe_2O_3 nanowire on the diameter of the wire (Adapted from Ref. [8]).

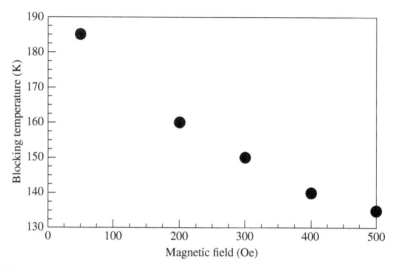

FIGURE 4.9   Dependence of the blocking temperature of Fe_2O_3 nanowires on the magnetic field strength (Adapted from Ref. [8]).

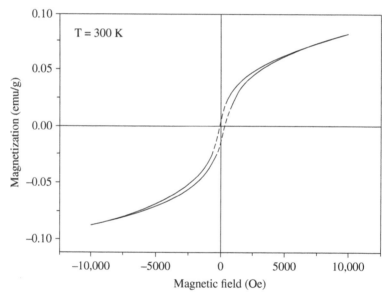

FIGURE 4.10   Magnetization versus magnetic field at room temperature for CuO nanorods (Adapted from Ref. [9]).

There are bulk materials that contain magnetic wires as subunits. Some examples are Li_2CuO that has CuO_2 chains. $CsMnCl_3 \cdot 2H_2O$ and $CuCl_2$ are other examples. Figure 4.11 shows the crystal structure of $CuCl_2$ showing the bulk solid contains chains of $CuCl_2$. These chains weakly interact with each other in the

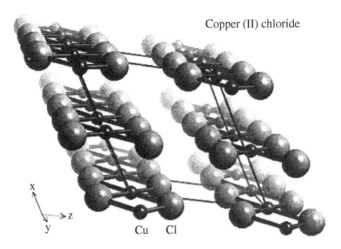

Copper (II) chloride

x
y z
Cu   Cl

FIGURE 4.11   Crystal structure of $CuCl_2$ showing the bulk solid contains $CuCl_2$ chains.

lattice and can be approximated as free chains. The structures are important because they allow a test of various theories of magnetism such as the Ising model, for which an exact solution can be obtained only for a one-dimensional magnetic system [10]. It has been shown to account fairly well for the temperature dependence of the susceptibility through the paramagnetic to antiferromagnetic transition in $CuCl_2$ [11].

Consider a lattice where each site has a spin S having values ±1 that can align either parallel or antiparallel to an applied magnetic field H. In the Ising model, the magnetic energy of the lattice is given by

$$E = J \sum_{i,j} S_i S_j - uH \sum_i S_i \qquad (4.1)$$

The first term is summed over the nearest neighbor pairs and the second over all spins of the lattice. For a ferromagnetic material, J is positive, and for an antiferromagnet, it is negative. For a one-dimensional system such as $CuCl_2$ in high magnetic fields, it has been shown that the Ising model gives the susceptibility as [11]

$$X = \left[\frac{N\mu^2}{kT} \right] \left[\exp\left(\frac{-4J}{kT} \right) + \left(\frac{\mu H}{kT} \right)^2 \right]^{-1/2} \qquad (4.2)$$

This equation has been shown to well describe the temperature dependence of the susceptibility in $CuCl_2$.

A transition from the paramagnetic state to a ferromagnetic or antiferromagnetic state is preceded by fluctuations of the magnetic moments. In effect, the randomly oriented magnetic moments have to undergo dynamical reorientation to achieve the

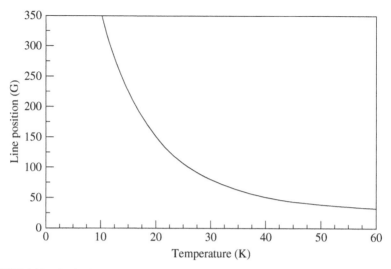

FIGURE 4.12 Shift of the magnetic field position of the EPR spectrum of Mn^{2+} in the paramagnetic phase of the linear chain material $CsMnCl_2 \cdot 2H_2O$ as the antiferromagnetic state is approached (Adapted from Ref. [12]).

aligned orientation in the magnetic phase. The details of the dynamics of the fluctuations depend on the dimensionality of the magnetic system. In one-dimensional systems, the fluctuations start at much higher temperatures above the transition temperature. The presence of these fluctuations has been observed in electron paramagnetic resonance (EPR) measurements. The fluctuations cause temperature-dependent deviations of the line width and g values as the transition temperature is approached from above. Further theoretical modeling of the fluctuation process is less complex than in higher dimensions. Thus, these one-dimensional systems make good materials to test the validity of various proposed models of magnetic fluctuations, and this has motivated a number of experimental studies using EPR. Generally, although not always, the EPR spectrum of a magnetic material is a broad line resulting from the magnetic interaction with the nearest neighbor magnetic atoms. The compound $CsMnCl_3 \cdot 2H_2O$ has a one-dimensional linear chain of $Cl-Mn^{2+}-Cl-Mn^{2+}$ that becomes antiferromagnetic at 4.89 K. Starting at about 20 K, there is a marked shift in the magnetic field position of the Mn resonance with lowering temperature when the magnetic field is parallel to the chain as shown in Figure 4.12 [12]. There is also a broadening of the line as the transition temperature is approached. These effects are due to fluctuations of the magnetic moments prior to the formation of the antiferromagnetic phase. In the context of the one-dimensional Heisenberg theoretical representation, models have been developed to predict how the fluctuations produce the line broadening and shifts in the field position with temperature. For example, the theory predicts the line width should increase as $1/T^{2.5}$, which is in reasonable agreement with experimental measurements.

4.2.3 Building One-Dimensional Magnetic Arrays One Atom at a Time

The scanning tunneling microscope (STM) uses a narrow tip to scan across the surface of the material about a nanometer above it. When a voltage is applied to the tip, electrons tunnel from the surface of the material and a current can be detected. If the tip is kept at a constant distance above the surface, then the current will vary as the tip scans the surface. The amount of detected current depends on the electron density at the surface of the material, and this will be higher where the atoms are located. Thus, mapping the current by scanning the tip over the surface produces an image of the atomic or molecular structure of the surface. An alternate mode of operation of the STM is to keep the current constant and monitor the deflection of the cantilever on which the tip is held. In this mode, the recorded cantilever deflections provide a map of the atomic structure of the surface. An illustration of the device is shown in Figure 4.13.

The STM has been used to build nanosized low-dimensional magnetic structures atom by atom on the surface of materials. An adsorbed atom is held on the surface by chemical bonds with the atoms of the surface. When such an atom is imaged in an STM, the tip has a trajectory of the type shown in Figure 4.14a. The separation between the tip and the adsorbed atom is such that any forces between them are small compared to the forces binding the atom to the surface, and the adsorbed atom will not be disturbed by the passage of the tip over it. If the tip is moved closer to the adsorbed atom, as shown in Figure 4.14b such that the interaction of the tip and the atom is greater than that between the atom and the surface, then the atom can be dragged along by the tip. At any point in the scan, the atom can be reattached to the surface by increasing the separation between the tip and the surface. In this way, adsorbed atoms can be rearranged on the surfaces of materials, and structures can be built on the surfaces atom by atom. The surface of the material has to be cooled to liquid helium temperatures in order to reduce thermal vibrations, which may cause the atoms to undergo thermally induced diffusion, thereby disturbing the arrangement of atoms being assembled. Thermal diffusion is a problem because this method of construction can only be carried out on materials in which the lateral or in-plane interaction of the adsorbed

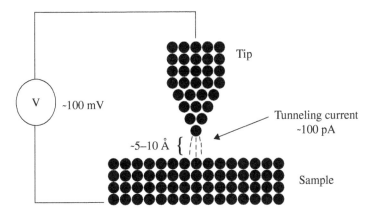

FIGURE 4.13   Illustration of a scanning tunneling microscope.

(a) Imaging mode

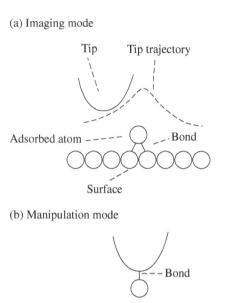

(b) Manipulation mode

FIGURE 4.14 Illustration of how the STM is used to build atomic structures on the surfaces of clean metals such as copper.

FIGURE 4.15 STM image of a linear chain of iron atoms built on the surface of copper using the STM (Adapted from Ref. [13]).

atom and the atoms of the surface is not too large. The manipulation also has to be done in an ultrahigh vacuum in order to keep the surface of the material clean. The method has been used to make low-dimensional arrays of iron atoms on the (111) surface of ultraclean copper. Figure 4.15 shows an STM image of a one-dimensional chain of seven iron atoms made by this process [13]. The linear chain has antiferromagnetic order. The dark images are the iron atoms with spin down, and the lighter images are iron atoms with spin up. It was observed that when the chain contained an even number of iron atoms, there is no antiferromagnetic order. For example, a chain consisting of six iron atoms had four atoms in the spin-down orientation and two in the spin-up orientation. These structures can provide an experimental test of the Ising model mentioned previously as it has an exact solution for a one-dimensional magnetic system. It was found that there were deviations of the magnetic field dependence of the magnetization from the predictions of the Ising model. The reason for this is not understood and presently under investigation.

4.3 MAGNETORESISTANCE IN BULK NANOSTRUCTURED MATERIALS

As discussed earlier, magnetoresistance is a phenomenon where the application of a dc magnetic field changes the resistance of a material. In 1988, giant magnetoresistance (GMR) was observed in materials synthetically fabricated by depositing on a substrate alternate layers of nanometer-thick ferromagnetic materials and nonferromagnetic metals. A schematic of the layered structure and the alternating orientation of the magnetization in the ferromagnetic layer is shown in Figure 4.16. The effect was first observed in films made of alternating layers of iron and chromium, but since then, other layered materials composed of alternating layers of cobalt and copper have been made, which display much higher magnetoresistive effects. Figure 4.17 shows the effect of a dc magnetic field on the resistance of the iron–chromium multilayered system [14]. The magnitude of the change in the resistance depends on the thickness of the iron layer, as shown in Figure 4.18, and it reaches a maximum at a thickness of 7 nm [14]. The effect occurs because of the dependence of electron scattering on the orientation of the electron spin with respect to the direction of magnetization. Electrons whose spins are not aligned along the direction of the magnetization M are scattered more strongly than those with their spins aligned along M. The application of a dc magnetic field parallel to the layers forces the magnetization of all the magnetic layers to be in the same direction. This causes the magnetizations pointing opposite to the direction of the applied magnetic field to flip. The conduction

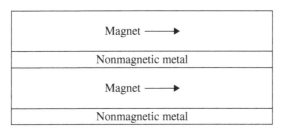

FIGURE 4.16 A structure for producing colossal magnetoresistance consisting of layers of nonmagnetic material alternating with oppositely magnetized (arrows) ferromagnetic layers.

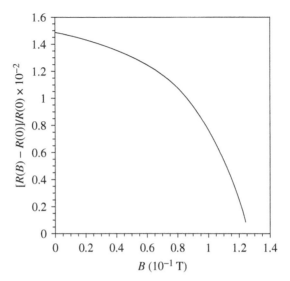

FIGURE 4.17 Dependence of the fractional change of the electrical resistance $R(B)$, relative to its value $R(0)$ in zero field, of a layered iron–chromium system on a magnetic field B applied parallel to the surface of the layers (From Ref. [14]. © IOP Publishing. Reproduced by permission of IOP Publishing. All rights reserved).

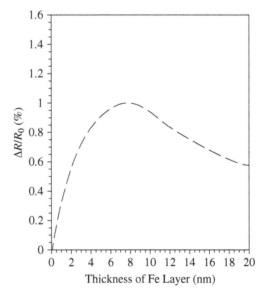

FIGURE 4.18 Dependence of the percent change of magnetoresistance $\Delta R/R_0$ on the thickness of the iron magnetic layer in a constant dc magnetic field for the Fe–Cr layered system (From Ref. [14]. © IOP Publishing. Reproduced by permission of IOP Publishing. All rights reserved).

electrons with spins aligned opposite to the magnetization are more strongly scattered at the metal–ferromagnet interface, and those aligned along the field direction are less strongly scattered. Because the two spin channels are in parallel, the lower resistance channel determines the resistance of the material.

The magnetoresistive effect in these layered materials is a sensitive detector of dc magnetic fields and is the basis for the development of a new more sensitive reading head for magnetic disks. Prior to this, magnetic storage devices have used induction coils to both induce an alignment of the magnetization in a small region of the tape (write mode) and to sense the alignment of a recorded area (read mode). The magnetoresistive reading head is considerably more sensitive than the inductive coil method.

A number of devices that depend on the difference in conductivity of conduction electrons depending on their spin orientation with respect to some axis such as an applied magnetic field have been developed. These devices depend on the fact that electrons with spin up and spin down with respect to the field direction have different transport properties in certain kinds of nanometer-thick layered materials. In metallic ferromagnetic materials, the spin-up and spin-down conduction electrons act as two independent families of charge carriers. The ferromagnetic exchange interaction causes a splitting of the spin-up electron conduction band from the spin-down electron conduction band. This results in different band structures and density of states at the Fermi level. The spin polarization (SP) is defined as the ratio of the difference of the spin-up minus the spin-down population to the total number of carriers at the Fermi level:

$$SP = \frac{N_{up} - N_{down}}{N_{up} + N_{down}} \quad (4.3)$$

If SP is not zero, there will be a different number of electrons in the two different spin channels. The scattering of the spin-up and spin-down electrons is different in these structures, which means the conductivity is different. Many of the structures where this difference is evident are layered materials of nanometer thickness.

Another place where spin is employed is in the hard drives of computers that store information based on the orientation of the magnetization of rod-shaped particles of nanometer thickness. A typical hard drive consists of an array of 1 µm long and about 70 nm wide particles. One device that has been developed to read hard drives is called a spin valve. It consists of nanometer-thick layers of materials, which collectively display GMR. The structure is illustrated in Figure 4.19. It consists of an antiferromagnetic layer on a ferromagnetic layer followed by a conducting layer and another ferromagnetic layer all deposited on a substrate. The ferromagnetic layer adjacent to the antiferromagnetic layer has its magnetization pinned parallel to the layer interface. The other magnetic layer referred to as the free layer can have the direction of its magnetization changed by a relatively small dc magnetic field. As the magnetization in the two layers changes from parallel to antiparallel, the resistance of the spin valve changes by 5–10%. Now, let us look at the behavior of some actual systems. Figure 4.20 shows the resistance versus dc

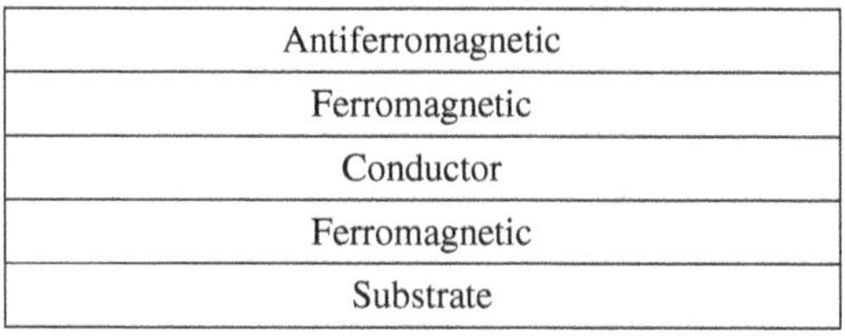

| Antiferromagnetic |
| Ferromagnetic |
| Conductor |
| Ferromagnetic |
| Substrate |

FIGURE 4.19   Illustration of the layered structure of a spin valve.

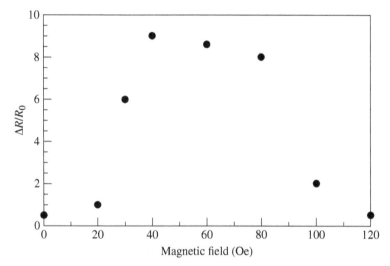

FIGURE 4.20   Fractional change in the resistance versus dc magnetic field of a layered structure of Co/Cu/Co where one of the Co layers has the direction of its magnetization pinned parallel to the interface between the layers (Reproduced with permission from Ref. [15]. Copyright 1991, AIP Publishing LLC).

magnetic field of a Co/Cu/Co structure where one of the ferromagnetic layers has the direction of its magnetization pinned by the exchange interaction with an adjacent antiferromagnetic material. The observed change in the resistance depends on the thickness of the layers. Figure 4.21 shows how the thickness of the Co layer in a spin valve structure having layers of composition Si/Co(x)/Cu(2.2)/NiFe(4.7)/FeMn(7.8)/Cu(1.5) affects the magnetoresistance [15]. The numbers in parenthesis are the thicknesses of the layers in nanometers. Notice that magnetoresistance is optimum at about a 10 nm thickness of the first cobalt layer. The thickness of the other layers also affects the magnetoresistance. Figure 4.22 shows how the saturation magnetization of the free Co layers depends on the layer thickness. These spin valve devices, because they are sensitive to small magnetic fields, are the basis of the reading of information on magnetic hard drives. It should be emphasized that these are truly nanometer-structured materials.

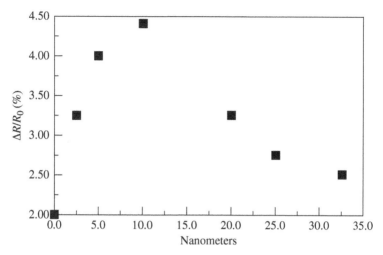

FIGURE 4.21 Magnitude of the relative change in resistance in a fixed applied magnetic field versus the thickness of the free magnetic layer for the spin valve structure discussed in the text (Reproduced with permission from Ref. [15]. Copyright 1991, AIP Publishing LLC).

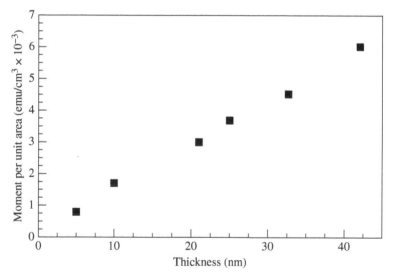

FIGURE 4.22 Plot of the magnetic moment of the same spin valve of Figure 4.19 versus the thickness of the free magnetic layer (Reproduced with permission from Ref. [15]. Copyright 1991, AIP Publishing LLC).

Magnetic tunneling junctions can also be used to read information on magnetic storage disks. In this application, the device is somewhat different than that shown in Figure 4.16. One of the ferromagnetic layers will have an antiferromagnetic layer deposited on it to pin the direction of the magnetization in that layer. The tunneling

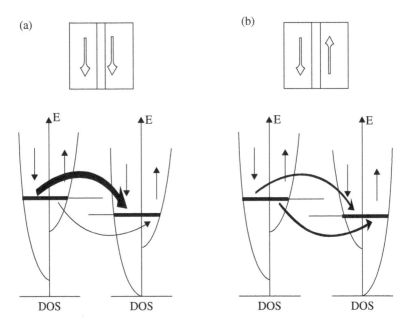

FIGURE 4.23   Illustration of the density of states and Fermi level for a magnetic tunnel junction in which the magnetizations are parallel (a) and antiparallel (b) with respect to each other.

current depends on the relative orientation of the magnetization of the two ferro-magnetic metal layers on either side of the insulating layer. Tunneling can only occur when there are available states on one side of the junction that can accept the electrons. This is accomplished by applying a voltage across the junction, which lowers the Fermi level on the right side of the junction. As discussed in Chapter 2, the density of states for electrons in a ferromagnetic material will be smaller for the spin-up electrons at the Fermi level. Thus, as illustrated in Figure 4.23a, more elec-trons will tunnel to the spin-down state on the right side when the magnetizations are parallel. There will still be tunneling to the electron spin-up state, but it will be smaller because the density of states is less. When the magnetizations are oppo-sitely aligned as shown in Figure 4.23b, there will still be tunneling but it will be much less and thus the resistance will be higher. The thickness of the arrows in the figure is proportional to the relative amount of electron tunneling. Figure 4.24 shows the tunneling resistance versus magnetic field at room temperature for a junction of $Co/Al_2O_3/NiFe$.

Magnetic tunnel junctions could also be used as information storage devices called magnetic random-access memory (MRAM). Because current can be turned on and off by small magnetic fields, magnetic tunnel junctions can be used to represent the one and zero of computer logic. Figure 4.25 illustrates a possible array of tunnel junctions that could be used as MRAM. It consists of layers of nanometer-thick

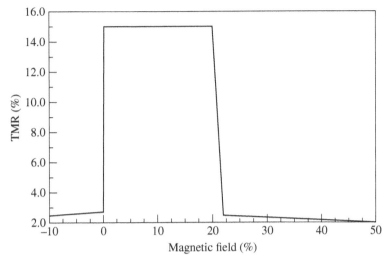

FIGURE 4.24 Change in resistance versus dc magnetic field at room temperature for a magnetic tunnel junction consisting of Co/Al$_2$O$_3$/NiFe nanometer-thick layers (Reproduced with permission from Nassar et al. [16]. Copyright 1998, AIP Publishing LLC).

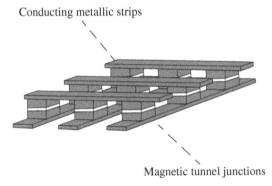

FIGURE 4.25 A proposed magnetic storage device based on magnetic tunnel junctions that exist at the overlapping regions where vertical and horizontal metallic strips cross over each other. The relative orientation of the spins in the two ferromagnetic layers stores the information bit.

metal strips overlaid on each other, as shown in the figure. Magnet tunnel junctions connect the horizontal and vertical strips where they cross over each other. Current flowing through the strips produces magnetic fields at the junctions. If the direction of current flow in the strips is such that magnetic fields on both sides of the junction are parallel, then current can flow between the horizontal and vertical strips. The relative orientation of the spins in the two ferromagnetic layers of the junction stores the information bit.

EXERCISES

4.1 Assume a two-dimensional lattice of iron particles having a square unit cell with an edge of 2.87Å. If a strip of this material of 1.44 nm in length and 0.57 nm in width were ferromagnetic, calculate the total magnetic moment of the lattice.

4.2 What is the magnetization of a Ni film that is 5 nm thick? Explain how your answer was obtained.

4.3 At what thickness does a Ni film become superparamagnetic?

4.4 At what grain size would a composite of Nd–B–Fe be superparamagnetic? Why is this smaller than a powder of the same material?

4.5 Calculate the total magnetic moment of the iron chain shown in Figure 4.15.

REFERENCES

1. K. Shafi et al. *J. Mater. Res.* 15, 332 (2000).
2. A. Manaf et al. *J. Magn. Magn. Mater.* 107, 360 (1991).
3. N. Chinnasamy et al. *J. Phys. Condens. Matter* 12, 7795 (2001).
4. A. Ruske, *Ann. Phys.* 2, 244 (1958).
5. B. Aktins, L. Tagirov, and F. Miknlov, *Magnetic Nanostructures*, p. 171, Springer-Verlag, New York, 2007.
6. J. Feng et al. *J. Am. Chem. Soc.* 133, 17832 (2011).
7. Y. Ma et al. *ACS Nano* 6, 1695 (2012).
8. L. Zhang and Y. Zhang, *J. Magn. Magn. Mater.* 321, L15 (2009).
9. H. Xiao et al. *Solid State Commun.* 141, 431 (2007).
10. J. M. Ziman, *Principles of the Theory of Solids*, p. 299, Cambridge University Press, Cambridge, 1964.
11. C. G. Barralough and C. F. No, *Trans. Faraday Soc.* 60, 836 (1964).
12. K. Nagata and Y. Tazuke, *J. Phys. Soc. Jpn.*, 32, 337 (1972).
13. A. A. Khajetoodians et al. *Nat. Phys.* 8, 497 (2012).
14. R. E. Camley and R. L. Stamps, *J. Phys. Condens. Matter* 5, 3727 (1993).
15. B. Dieny et al. *J. Appl. Phys.* 69, 4774 (1991).
16. J. Nassar et al. *Appl. Phys. Lett.* 73, 689 (1998).

5

MAGNETISM IN CARBON AND BORON NITRIDE NANOSTRUCTURES

5.1 CARBON NANOSTRUCTURES

5.1.1 Fullerene, C_{60}

The C_{60} molecule was synthesized in 1985. The structure of the molecule is shown in Figure 5.1. It has 12 pentagonal (5 sided) and 20 hexagonal (6 sided) faces symmetrically arrayed to form a molecular ball. As can be seen from Figure 5.1, every pentagon is surrounded by five hexagons. Each carbon atom is bonded to three other carbon atoms by sp^2 hybrid orbitals as in graphite. However, because of the curvature of the surface, there is some admixture of sp^3 orbitals. The average C–C bond length of 1.44A is quite close to that of graphite. Two single bonds are located along the edge of a pentagon having a bond length of 1.46A. The double bonds have a length of 1.40A. The structure has a relatively high electron affinity, and an interesting result of this is that the anion of C_{60} is slightly more stable than neutral C_{60} by about 0.74eV as determined from molecular orbital calculations. Generally, the C_{60} anion is present in any batches of C_{60} and can be observed by electron paramagnetic resonance (EPR). The high electron affinity makes C_{60} a good radical scavenger, which means it can attract and bond to free radicals. The C_{60} molecule has many normal modes of vibration. The most intense Raman absorption at $1469\,cm^{-1}$ is called the pentagonal pinch mode. The motion of the atoms in this vibration involves a tangential displacement of the carbon atoms along with a contraction of the pentagonal rings and an expansion of the hexagonal rings.

When solid C_{60} is subjected to UV light in vacuum, it polymerizes into short polymers referred to as oligomers. The evidence for this is obtained by mass spectrometry. The classical gas mass spectrometer, sketched in Figure 5.2, ionizes the nanoparticles to form positive ions by impact from electrons emitted by the electron

Physics of Magnetic Nanostructures, First Edition. Frank J. Owens.
© 2015 John Wiley & Sons, Inc. Published 2015 by John Wiley & Sons, Inc.

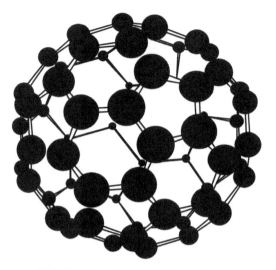

FIGURE 5.1   Structure of C_{60} molecule.

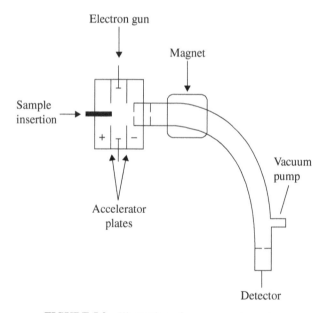

FIGURE 5.2   Illustration of a mass spectrometer.

gun in the ionization chamber of the ion source. The newly formed ions are accelerated through the potential drop in a voltage, V, between the accelerator plates, then
focused by lenses, and collimated by slits during their transit to the mass analyzer.
The magnetic field B of the mass analyzer, oriented normal to the page, exerts the
force $F = qvB$ that bends the ion beam through an angle of $90°$ at the radius r, after

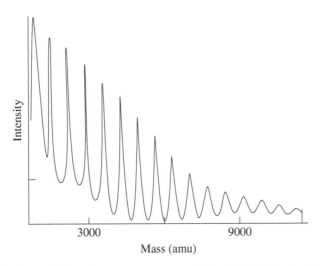

FIGURE 5.3 Laser deabsorption mass spectrum of C_{60} subjected to photolysis, showing formation of oligomers (Adapted from Ref. [1]).

which they are detected at the ion collector. The particle has a charge q and velocity v. The mass m to charge q ratio is given by the expression

$$\frac{m}{q} = \frac{B^2 r^2}{2v} \tag{5.1}$$

The bending radius r is ordinarily fixed in a particular instrument, so either the magnetic field B or the accelerating voltage V can be scanned to focus the ions of various masses at the detector. The charge q of the nanoion is ordinarily known, so in practice, it is the mass m that is determined. Figure 5.3 shows the mass spectrum of a film of C_{60} that has been UV irradiated for 12 h [1]. The spectrum is obtained by deabsorbing the material from the film in a vacuum using a pulsed laser. The deabsorbed material is then sent to the analyzer of the mass spectrometer. The spectrum consists of a series of peaks occurring at mass positions that are multiples of atomic mass 720, which is the atomic mass of C_{60}. The first largest peak is from molecular C_{60}. The second largest peak occurs at mass 1440, which corresponds to the formation of a dimer. The structure of the dimer is shown in Figure 5.4.

When the photopolymerization occurs, the Raman spectra of the pentagonal pinch mode shifts from 1469 to 1459 cm⁻¹ [2]. When the photopolymerized material is heated above about 125°C, the intensity of the 1459 cm⁻¹ mode begins to decrease, and the intensity of the 1469 cm⁻¹ mode begins to increase indicating the dimer is dissociating into two fullerene molecules. This means the dimer phase is not stable much above 150°C. Figure 5.5 shows the temperature dependence of the two lines, which allows a determination of the kinetics of the return of the dimer phase to the monomer.

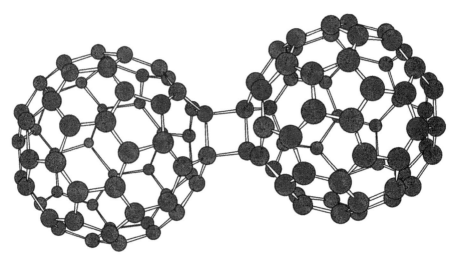

FIGURE 5.4   Structure of C_{60} dimer.

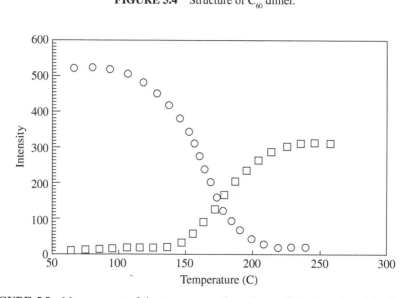

FIGURE 5.5   Measurement of the temperature dependence of the intensity of the Raman spectra of the pentagonal pinch mode of the C_{60} dimer at $1459\,cm^{-1}$ (O) showing a decrease as the temperature increases with a concomitant increase in the pinch mode intensity of an isolated C_{60} at $1469\,cm^{-1}$ ($\square$) (Adapted from Ref. [2]).

5.1.2   Carbon and Boron Nitride Nanotubes

Single-walled carbon nanotubes (SWNTs) are tubes made of carbon having a graphite-like structure. They can be envisioned as single sheets of graphite rolled into tubes. Figure 5.6a shows a sheet of graphite. If a tube is formed by folding the tube about an axis parallel to a C–C bond, a zigzag tube is formed shown in Figure 5.6b.

(a)　　　　　　　　　(b)　　　　　　　　　(c)

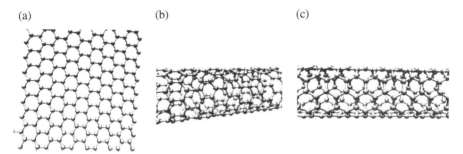

FIGURE 5.6 (a) Illustration of a sheet of graphene, (b) a zigzag carbon nanotube, and (c) an armchair carbon nanotube.

If it is folded about an axis perpendicular to a C–C bond, an armchair structure is produced shown in Figure 5.6c. The difference in the surface geometry of SWNTs is referred to as chirality. Typically, SWNTs are hundreds of microns long and have diameters of 5–20Å, making them the thinnest known fibers. They have tensile strengths 20 times that of steel. Tensile strength is the force per unit area needed to pull a tube apart. SWNTs can be metallic or semiconducting depending on their chirality and have thermal conductivities as high as diamond. Because of these properties, carbon nanotubes have large potential to improve the properties of materials such as strength and electrical conductivity by fabricating composites of the SWNTs with various materials.

There are a number of challenges to developing SWNTs into electronic and magnetic devices. The band gap of SWNTs depends on tube diameter and chirality. Synthesis generally produces a mix of semiconducting and metallic SWNTs having different chiralities, which require development of methods to separate the different types. Further synthesis uses magnetic nanoparticles such as Fe, Ni, and Co that remain included in the materials. It is difficult to remove all the magnetic particles from the tubes. Their presence will mask any intrinsic magnetism that may be present. Multiwalled tubes, which are tubes nested inside tubes, as illustrated in Figure 5.7, do not require nanoparticle catalysts and therefore could be candidates to investigate the possibility of intrinsic magnetism.

There is another carbon nanotube structure called a single-walled carbon nanohorn. It is a tubular carbon structure having caps at the end with a diameter approximately the same size as C_{60}. Because this structure can be synthesized without the need for metal catalysts, it also is a candidate for investigating the possibility of ferromagnetism.

Boron nitride (BN) nanotubes, shown in Figure 5.8, which have the same structure as SWNTs, provide an interesting alternative for developing electronic devices because their electronic properties, such as the band gap, do not depend on tube diameter, tube length, or chirality. Also, they do not require magnetic nanoparticles in order to catalyze the growth. Further, they are more resistant to oxidation compared to SWNTs. BN nanotubes have one major disadvantage: they are not semiconductors having band gaps greater than 5.5 eV. However, calculations,

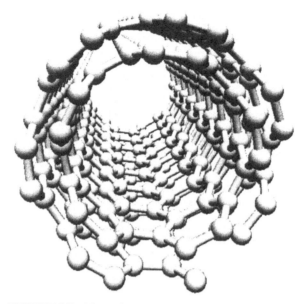

FIGURE 5.7 Illustration of a multiwalled carbon nanotube.

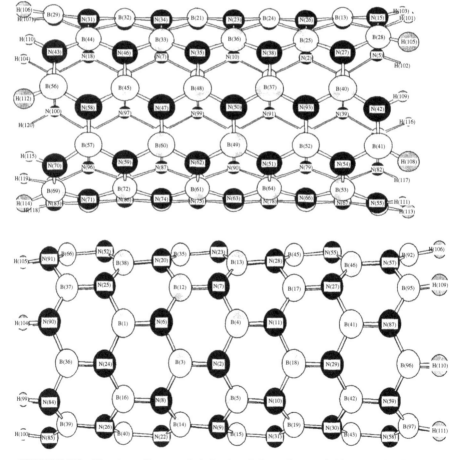

FIGURE 5.8 Structure of an armchair (top) and zigzag boron nitride nanotube (bottom).

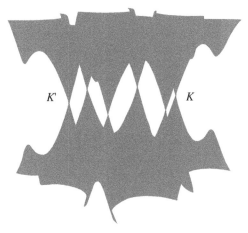

FIGURE 5.9 Energy versus K vector of graphene in the valence band region showing the linear dependence of the energy on the K vector near $K=0$.

which will be discussed later, indicate it is possible with appropriate doping or changing the ratio of B to N in the tubes to reduce the band gap to values that could make them semiconductors.

5.1.3 Graphene

Graphene is a two-dimensional array of carbon atoms having the same structure as carbon in the planes of graphite. Its structure is illustrated in Figure 5.6a. Graphene has unique electronic properties because the dependence of the band energies on the wave vector is linear near $K=0$ resulting in the conduction electrons having zero effective mass. This means they behave like zero-mass relativistic fermions. Figure 5.9 illustrates the dependence of the structure of the conduction band and valence band on the wave vector, K, showing the linear dependence on the wave vector near $K=0$. Field effect transistors have been demonstrated using graphene nanoribbons. Such graphene transistors could have very high switching times, since electrons move through graphene at extremely high speeds, 10–100 times faster than in present-day silicon chips.

5.2 EXPERIMENTAL OBSERVATIONS OF MAGNETISM IN CARBON AND BORON NITRIDE NANOSTRUCTURES

5.2.1 Magnetism in C_{60}

The C_{60} molecule has a very high electron affinity, which means it strongly attracts electrons. On the other hand, the molecule $C_2N_4(CH_3)_8$ or tetrakis(dimethylamino) ethylene (TDAE) is a strong electron donor, which means it readily gives its electron to another molecule. When C_{60} and TDAE are dissolved in a solvent of benzene and toluene, a precipitate of a new material forms that consists of the TDAE molecule complexed with C_{60}. This 1:1 complex has a monoclinic crystal structure. A measured large increase in the susceptibility at 16 K indicated this material is ferromagnetic [3].

When C_{60} is irradiated with UV light in vacuum, it forms short polymers of nanometer length as discussed earlier. It was found that when the photolysis is done in the presence of oxygen or air, a ferromagnetic phase is formed. Figure 5.10a shows a measurement of the magnetization versus dc magnetic field using a SQUID magnetometer at a number of temperatures in photolyzed C_{60} showing clear evidence of the existence of ferromagnetism at 300 K [4, 5]. Figure 5.10b shows a plot of the temperature dependence of the magnetization in a 300 G magnetic field. The temperature dependence of the magnetization $M(T)$ generally obeys the Bloch equation, Equation 2.12 of Chapter 2. The line through the data in Figure 5.10b is a fit to this equation. The Curie temperature estimated from the fit, where $M(T)$ becomes zero, is 1000 K, which is probably an overestimate. Interestingly, the ferromagnetism can be experimentally observed up to 800 K, which means the oligomers may not be the source of the ferromagnetism. The polymers returned to the monomer state in the vicinity of 150°C as seen by the Raman data in Figure 5.5. This Curie temperature is the highest temperature ever observed for an organic ferromagnetic material. An interesting feature of the result is that the ferromagnetic phase is not produced when the photolysis is done in a vacuum. This suggests that the photolysis in air involves a photodissociation of O_2 with a subsequent attachment of atomic oxygen to the C_{60} molecules.

There are a number of reports of ferromagnetism in halogenated C_{60}. C_{60} subjected to a heat treatment in the presence of iodine was shown to be ferromagnetic below 60 K. Doping with a mix of iodine and bromine produced a material that was ferromagnetic below 30 K. Fluorine-doped C_{60} was shown to be ferromagnetic having a Curie temperature above room temperature [6]. Because the fluorine-doped material has a Curie temperature above room temperature, the evidence for it will be examined in some detail.

The C_{60} was exposed to fluorine by embedding it in a fluorine-rich polymer, polytrifluorochloroethylene (PTFCE), and decomposing the polymer–C_{60} mix at high temperature to produce fluorine. In a typical process, 0.084 g of PTFCE was mixed with 0.040 g of C_{60}. The mixture was heated to 170°C for two minutes and then rapidly quenched to room temperature. Heating to higher temperatures and or longer times did not produce ferromagnetism. Figure 5.11 shows a plot of the magnetic field dependence of the magnetization measured by SQUID magnetometry. This field-dependent magnetization is not observed in the separated starting materials subjected to the same treatment. Figure 5.12 presents the temperature dependence of the magnetization above room temperature, measured in a 3000 G field and normalized to its value at 300 K. A fit of this data to the Bloch equation yields a value of A of 4.1246×10^{-5}. Using this the temperature at which $M(T)$ is zero, the Curie temperature is estimated to be 837 K. This is very close to the estimated Curie temperature produced by subjecting C_{60} to UV light in the presence of oxygen.

The material was heated above its melting point, and a 4000 G magnetic field applied. The material was then cooled below the melting temperature in the magnetic field. This aligns and locks in the direction of maximum magnetization parallel to the

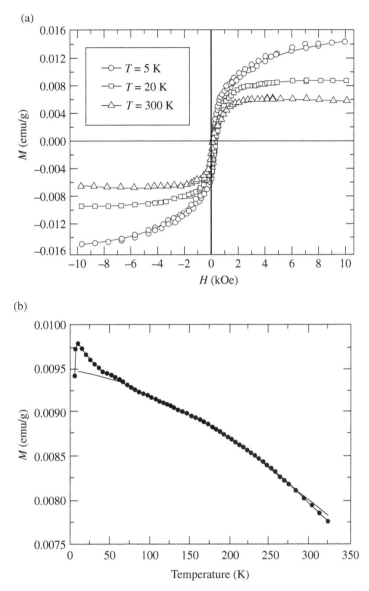

FIGURE 5.10   (a) Magnetic field dependence of magnetization of photolyzed C_{60} at three temperatures and (b) temperature dependence of the magnetization (Adapted from Ref. [4]).

direction of the cooling magnetic field. Figure 5.13 shows the ferromagnetic resonance (FMR) spectra for the sample oriented perpendicular and parallel to the direction of the cooling field.

The spectrum displays the characteristic temperature dependence of an FMR signal. Figure 5.14 shows a plot of the temperature dependence of the field position for the sample oriented perpendicular to the direction of the cooling field showing a

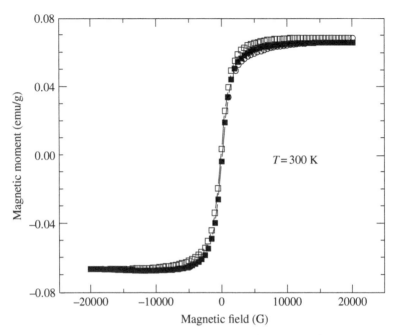

FIGURE 5.11 SQUID magnetometry measurement of Magnetic field dependence of magnetization of fluorinated C_{60} at room temperature (Courtesy of K. V. Rao, Royal Institute of Technology, Sweden).

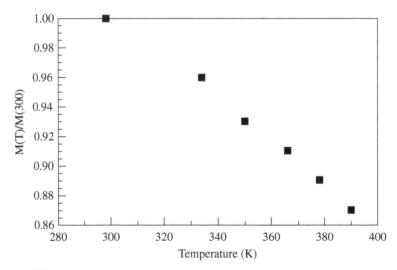

FIGURE 5.12 Temperature dependence of the magnetization of fluorinated C_{60} in a 3000 G magnetic field (Adapted from Ref. [6]).

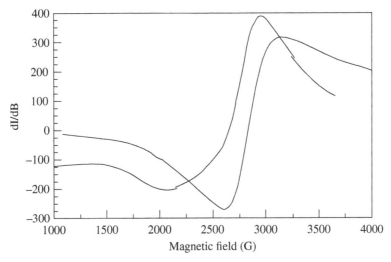

FIGURE 5.13   Ferromagnetic resonance spectra of the C_{60}—PTFCE composite parallel and perpendicular to the direction of an applied magnetic field in which the polymer was solidified (Adapted from Ref. [6]).

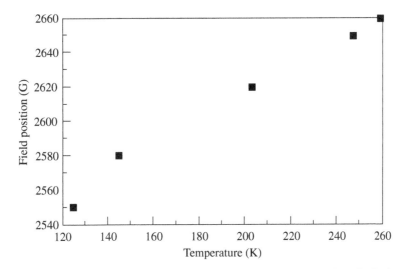

FIGURE 5.14   Temperature dependence of the magnetic field position of the FMR signal of the C_{60} PTFCE composite parallel to the direction of the cooling field (Adapted from Ref. [6]).

pronounced decrease in the field position with decreasing temperature. Figure 5.15 shows a plot of the line width as a function of decreasing temperature showing a marked broadening as the temperature is lowered. The data in Figures 5.14 and 5.15 confirm that the signal is an FMR signal. The previous data clearly indicate the presence of ferromagnetism in the fluorinated C_{60} having a Curie temperature well above room temperature.

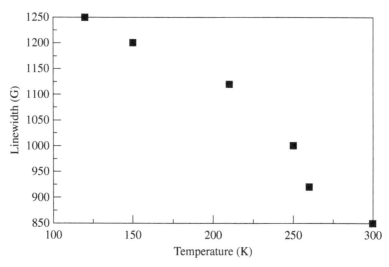

FIGURE 5.15   Temperature dependence of the line width of the FMR spectrum of the C_{60}
PTFCE composite parallel to the direction of the aligning field (Adapted from Ref. [6]).

The detailed structure of the ferromagnetic phase is unclear. There have been a
number of theoretical calculations that provide some insight. The most widely used
method to theoretically treat carbon and boron nitride nanostructures is density
functional theory (DFT). A brief overview of the theory is presented in Appendix C.
Calculations have been made to determine if the necessary spin arises from carbon
vacancies in the C_{60} framework of the polymers [7]. The calculations removed one
carbon atom in each C_{60} unit of the linear oligomer. The vacancies gave rise to a net
spin indicating a magnetic ground state. However, it is difficult to see how such
defects can be created in the room-temperature photogenerated magnetic oligomers
or in the fluorinated polymers that are produced at a relatively low temperature. This
is particularly true since the binding energy per carbon atom is 7.4 eV, which accounts
for the high stability of C_{60} [8]. Another possibility indicated by DFT molecular
orbital calculations on R–C_{60}=C_{60}–R dimers, where R could be H, O, F, or OH, is the
prediction that the triplet state has a lower energy than the singlet state [9]. The
calculations indicated that the triplet state had a lower energy than the singlet state by
0.51–0.55 eV depending on the nature of R. This suggests a possible origin for the
needed spin to produce ferromagnetism. However, calculations of the energy
necessary to dissociate the dimers into monomers indicate that the dimers would not
be stable at the high temperatures at which the ferromagnetism is observed. The calculated bond dissociation energies ranged from 1.10 to 0.28 eV. The lower values
were for negatively charged dimers such as (O–C_{60}=C_{60}–O)⁻ and (C_{60}=C_{60}O)⁻.
    Another possibility is an F atom bonds to the C_{60} molecule in the C_{60} crystal lattice
that has a face-centered cubic (FCC) structure. Since fluorine has nine electrons, a
single F atom bonded to C_{60} would provide C_{60} with an unpaired electron necessary
for ferromagnetism. It was found that the calculated energy to remove an F atom

from a single C_{60} was quite high having a value Of 4.36 eV. These results suggest that the most likely structure for the observed ferromagnetism is a lattice consisting of a C_{60} functionalized with a single atom or group having an unpaired electron.

5.2.2   Ferromagnetism in Carbon and Boron Nitride Nanotubes

While there are many theoretical predictions of the possibility of ferromagnetism in tubes having vacancies or doped with various atoms such as boron or nitrogen, there are no substantiated reports of intrinsic ferromagnetism in SWNTs. One of the problems is the presence of the magnetic catalysts necessary for synthesis would mask any intrinsic ferromagnetism.

Multiwalled carbon nanotubes grown inside alumina templates were shown to contain many defects. When subjected to hydrogen gas, the tubes absorbed considerable hydrogen and were shown to be ferromagnetic having a Curie temperature near 1000 K [10].

In another experiment, it was shown that multiwalled carbon nanotubes can become magnetized when they are in contact with a magnetic material [11]. The magnetism was observed using a magnetic force microscope. Ferromagnetism above room temperature has been observed in double-walled carbon nanotubes that were subjected to an acid treatment to remove the iron catalyst. The magnetization at room temperature was in the order of 2.2 emu/g, which was considerably greater than the estimated value of the magnetization from the residual iron catalyst, which was determined to be 0.022 emu/g. The Curie temperature was well above room temperature. The ferromagnetism was attributed to adsorbed hydrogen on the tubes due to the acid purification process.

Capped nanotubes, shown in Figure 5.16, which do not require metal catalysts for synthesis, show evidence for ferromagnetism when exposed to oxygen at high pressures [12]. Because the tubes were closed at both ends, they were heated to between 400 and 500°C, which opens the ends of the tubes allowing oxygen to penetrate the interior of the tubes. Magnetism was observed at low temperature, in the vicinity of 50 K, in the treated material. The magnetism is attributed to the presence of solid oxygen inside the tubes or absorbed on the surface of the tubes. There have been no experimental reports of ferromagnetism in boron nitride or boron nanotubes.

While there have been only a few experimental hints of magnetism in SWNTs, there have been many theoretical predictions of the possibility of magnetism induced by doping, sidewall functionalization, or defects. Here, we will discuss a few representative examples rather than discuss every single report. It has been shown that to get reliable predictions of properties, the nanotubes must have at least 70 atoms. DFT has been used to investigate the magnetic properties of armchair and zigzag SWNTs having different diameters and carbon vacancies [13]. It is predicted that metallic SWNTs having a vacancy could be ferromagnetic [14]. However, semiconducting tubes were not predicted to be ferromagnetic. The effect of doping armchair and zigzag SWNTs with two boron or two nitrogen atoms was investigated [15, 16]. The structures were optimized in the singlet and triplet states using molecular orbital theory. The triplet state of the zigzag tubes, but not the armchair tubes, was predicted

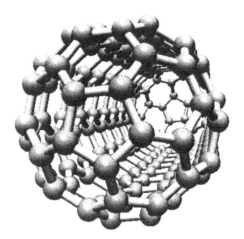

FIGURE 5.16   Illustration of a capped carbon nanotube.

to have a lower energy than the singlet state. For the zigzag tubes, the density of states at the top of the valence band is predicted to be greater for the spin-down state compared to the spin-up state indicating the doped tubes could be ferromagnetic. For example, in the case of the tubes doped with two borons, the triplet state is predicted to be 4.23 eV below the singlet.

The possibility of ferromagnetism in BNNTS is of much interest because their synthesis does not require magnetic catalysts that would mask the presence of intrinsic ferromagnetism in the tubes. There have been a number of calculations that predict ferromagnetism in various modifications of BN tubes.

Zigzag BN tubes having a boron vacancy were predicted to be ferromagnetic using DFT [17, 18]. However, armchair tubes were not predicted to be ferromagnetic. Calculations of carbon-doped BN zigzag and armchair tubes using DFT and the local spin density approximation predicted that both kinds of tubes would be ferromagnetic [19]. However, the band gaps were predicted to be large indicating the tubes would be ferromagnetic insulators.

5.2.3 Magnetism in Graphene

The first prerequisite for the existence of ferromagnetism is the existence of unpaired electron. When the width of graphene sheets, shown in Figure 5.17a and b, is reduced to a few nanometers, a number of new and interesting properties emerge. There are two kinds of nanoribbons shown in Figure 5.17, zigzag in 5.17a and armchair in 5.17b. The zigzag and armchair designations can be distinguished by the orientations of the C–C bonds at the edge of the ribbons. In the case of zigzag ribbons, the edge C–C bonds are parallel to the length of the ribbon, whereas in the armchair ribbons, there are no edge bonds parallel to the length of the ribbon. Consider the zigzag ribbon shown in Figure 5.17; as the length is gradually increased through the series $C_{28}H_{14}$, $C_{35}H_{15}$, $C_{42}H_{18}$, $C_{49}H_{19}$, $C_{56}H_{22}$, $C_{63}H_{23}$ to $C_{70}H_{26}$, the number of carbon atoms

(a)

(b)

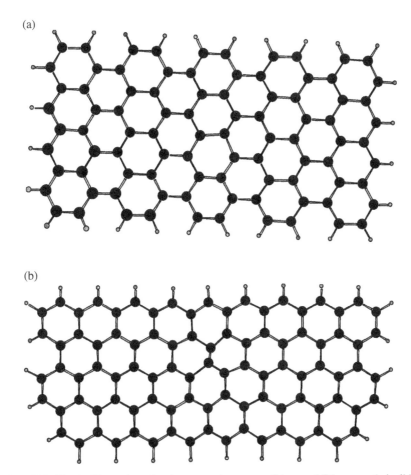

FIGURE 5.17   (a) Illustration of a zigzag graphene nanoribbon and (b) an armchair ribbon.

oscillates between odd and even. The ribbons having an odd number of carbons have an unpaired electron, a prerequisite for the existence of ferromagnetism. On the other hand, oscillation between odd and even carbon number does not occur in armchair ribbons. Thus, there are carbon nanoribbons that can have an unpaired electron without the presence of defects.

Ferromagnetism has been observed in a two-dimensional carbon nanosheet similar to that illustrated in Figure 5.6a [20]. The sheets were fabricated using an inductively coupled radio-frequency (RF) technique. A chamber containing an RF source is used to generate a plasma from a mixture of argon and methane gas. The graphite layers are formed on a silicon substrate from carbon atoms that are products of the decomposition of methane, CH_4. The substrate was held at a constant temperature of 400°C. Various spectroscopic techniques such as atomic force microscopy, x-ray diffraction, and Raman spectroscopy were used to verify the existence of the one-dimensional graphite layer. Only the material made at the longest deposition time of 120 min shows evidence of ferromagnetism. Raman spectroscopy was used to characterize the sample made at different deposition times. The spectra consist of two strong lines at 1588 cm^{-1} referred to as the G mode and 1322 cm^{-1} designated as the D mode. The G mode is due to a stretching vibration of the carbon rings parallel to the graphite plane. The D mode is due to a breathing vibration of the sp^2 atoms in the ring, and its intensity is sensitive to the number of defects in the graphite plane. The ratio of the intensity I_g/I_d can be used to assess the defect content in the graphite plane. The smaller the ratio, the greater the defect content. The material made with the two lowest deposition times had a I_g/I_d ratio close to one indicating a highly defected structure and low degree of graphite formation. The material with the longest deposition time has a ratio of 2.7 indicating a low defect content.

There have been other observations of ferromagnetism in graphene-like structures. For example, graphene made from graphene oxide that can be chemically synthesized has been shown to display ferromagnetism having a maximum magnetization of 0.010 emu/g [21]. The origin of the ferromagnetism has not been determined but is likely due to the presence of some kind of defects. One possibility is that the hydrogen released in the decomposition of methane bonds to the surface of the graphene plane. There have been theoretical predictions that surface hydrogenated graphene could be ferromagnetic.

One defect that is not likely to cause the ferromagnetism is a carbon vacancy. Carbon vacancies have been produced in graphene by proton irradiation, and the resulting material has been shown to be paramagnetic with the spin ½ entities localized around the vacancy [22]. However, it was not possible to achieve more than one magnetic moment per 1000 carbon atoms. Graphene with a greater vacancy concentration would likely be unstable. DFT calculations of the minimum-energy structure of graphene ribbons having a single vacancy indicate the ribbons are distorted from planarity [23]. Since ferromagnetism results from exchange interaction between unpaired spins, which requires that the distance between the spins be relatively small, it is unlikely that carbon vacancies in the graphene plane can be the cause of the ferromagnetism. It should be emphasized that the observed magnetizations in all observations are quite small and most of the observations have not been replicated to date.

There is much more research needed to establish the existence of ferromagnetism in graphene and determine its origin.

As in the case of SWNTs, there have been many predictions of ferromagnetism in graphene and two-dimensional boron nitride sheets. Here, we will highlight a few representative examples. Generally, most calculations obtain the minimum-energy structure using spin-polarized DFT employing the generalized gradient approximation. Some calculations employ periodic boundary conditions that enable the band gap to be determined as a function of the K vector in the Brillouin zone. Those that do not use periodic boundary conditions can only obtain the band gap at $K=0$, provided the calculation is done on a structure having a sufficiently large number of constituents.

It has been predicted that a vacancy in graphene is spin polarized having a magnetic moment of 1.04 u$_B$ [24]. Each of the three neighboring carbon atoms has

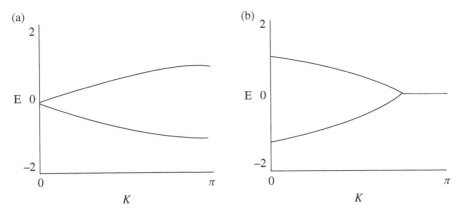

FIGURE 5.18   Calculation of the dependence of the highest occupied molecular orbital and the lowest unoccupied molecular orbital versus K vector for a zigzag ribbon (b) and an armchair ribbon (a) (Adapted from Ref. [8]).

one sp^2 dangling bond. Ferromagnetism has also been predicted if two of the dangling bonds have hydrogen attached to them. Graphene nanoribbons doped with nitrogen and boron atoms have been predicted to have triplet ground states [8]. The density of states at the valence band was predicted to be larger for the spin-down state compared to the spin-up state, suggesting the possibility of ferromagnetism.

The energy levels of a solid depend on the K vector, which in the free electron model of metals is given by Equation 1.7 where $K=n/2L$ for the one-dimensional lattice. In order to fully characterize the energy levels of a solid, their dependence on the K vector is needed. The tight-binding model outlined in Appendix D can be used to calculate the K dependence of the energies of graphene [25]. The calculations reveal some unusual features, which bear on the possibility of ferromagnetism in graphene. Figure 5.18a and b shows the calculated dependence of the energy of the highest occupied molecular orbital (HOMO) and the lowest unoccupied molecular orbital (LUMO) for a zigzag and armchair ribbon having a width of four carbon rings as shown in Figure 5.17. The energies are normalized to t, which is the nearest neighbor overlap integral in the tight-binding model. The results for the zigzag ribbon show an unusual coincidence of the energy levels of the HOMO and the LUMO near $K=\pi$. This degeneracy is not predicted for graphite or the armchair ribbons. It is found that in this region where the energies are coincident, charge is localized on the zigzag edges. These states are referred to as edge states. Theory also predicts that there are magnetic moments associated with these edge states and that these moments are ordered ferromagnetically on one side of the ribbon and antiferromagnetically on the opposite side. This means the net magnetic moment of the ribbon is zero for a ribbon having an equal number of edge states on each side.

There is some preliminary experimental evidence for the edge states in graphene ribbons. Near-edge x-ray absorption fine structure (NEXAFS) spectroscopy has been used to investigate this issue [26]. In this experiment, an x-ray photon excites an electron from the core carbon first level, producing a photoelectron emission. The energy

(a)

(b)

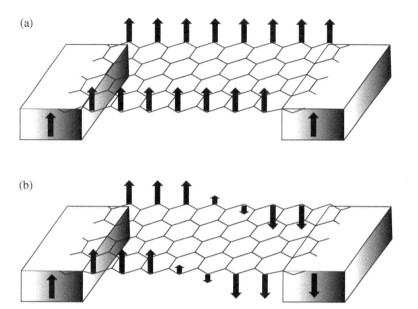

FIGURE 5.19   Illustration of a magnetoresistive device based on a zigzag graphene nanorib-bon having edge states (Reproduced with permission from Ref. [28] © IOP Publishing).

of the emitted electron is measured. In graphite, there is a peak at 285.5 eV correspond-ing to a transition from the carbon first level to the LUMO state. In graphene, an addi-tional small peak was observed on the low-energy side of the graphite peak and attributed to spins at the edges.

The graphene was synthesized by a chemical vapor deposition method, and the samples were then annealed at different temperatures. The new peak was only observed in the samples annealed at 1000 and 1500°C. A narrow EPR signal was also observed in the same samples and attributed to spins at the edges. Both the line width and g value decreased as the temperature was lowered, which was attributed to a strong coupling of the spins to the conduction carriers. These effects could also be due to the onset of ferromagnetic order of the edge spins.

Another approach to producing spins at the edges is to remove hydrogen atoms from the edges that could leave unpaired electron at the edges. Theoretical DFT cal-culations have in fact predicted that ribbons with half the hydrogen atoms removed could be ferromagnetic semiconductors [27].

If these edge states exist, they could provide a basis for some interesting magnetic devices. One proposal that has been suggested is they could be made into magneto-resistive devices. Magnetoresistive devices, which have been discussed in Chapter 4, consist of two ferromagnetic layers separated by a metallic nonferromagnetic layer. Zigzag graphene nanoribbons having two metallic contacts on each end should display magnetoresistive effects. The concept of the device is illustrated in Figure 5.19 [28]. The low resistive state occurs when the magnetic moment of the two ferromag-netic metal contacts are parallel as illustrated in Figure 5.19a. This forces the spins at the edges to be parallel to the magnetic moment of the metal contacts. When the magnetic moments of the two metal contacts are antiparallel, the edge state spins align parallel to the magnetic moment of the closest metal contact. In this configuration, the resistance is highest.

EXERCISES

5.1 If an OH is attached to the open end of a carbon nanotube and to the sidewall that has no carbon vacancies, which bond would be stronger and why?

5.2 The dispersion relationship for graphene in the tight-binding model has been shown to be 3/2talKl where *a* is the lattice constant and *t* is an overlap integral. Calculate the effective mass of graphene. How is the conductivity affected by the result? What are the implications for possible applications of graphene?

5.3 It has been reported that a graphene-like structure can be made of silicon. What advantages would such structures have compared to graphene?

5.4 Why is it unlikely that doping or creating vacancies in graphene would lead to ferromagnetism?

5.5 Some evidence has been presented for the existence of a smaller fullerene such as C_{20}. If these were to form oligomers and they were observed by mass spectrometry, at what masses would the first three lines be observed?

REFERENCES

1. A. M. Rao, et al. *Science* 259, 955 (1993).
2. Y. Wang, et al. *Chem. Phys. Lett.* 211, 341 (1993).
3. P. M. Allemand, et al. *Science* 253, 301 (1991).
4. F. J. Owens, Z. Iqbal, L. Belova and K. V. Rao, *Phys. Rev.* B69, 033403 (2004).
5. Y. Murakami and H. Suematsu, *Pure Appl. Chem.* 68, 1463 (1996).
6. F. J. Owens and R. Patel, *J. Mater.* 261304 (2013).
7. A. N. Andriotis, M. Menon, R. M. Sheetz and L. Chernozatonskii, *Phys. Rev. Lett.* 90, 026801 (2003).
8. S. Satio and A. Oshiyama, *Phys. Rev.* B44, 11532 (1991).
9. M. Miller and F. J. Owens, *Chem. Phys.* 394, 52 (2012).
10. A. L. Frieman, *Phys. Rev.* B81, 115461 (2010).
11. O. Cespedes, *J. Phys. Condens. Matter* 16, L155 (2004).
12. S. Bandow, T. Yamaguchi and S. Iijima, *Chem. Phys. Lett.* 401, 380 (2005).
13. Y. Ma, P. O. Lehtinen, A. S. Foster and R. M. Nieminen, *New J. Phys.* 6, 68 (2004).
14. W. Orellana and P. Fuentealba, *Surf. Sci.*, 600, 4305 (2006).
15. F. J. Owens, *Mater. Lett.* 61, 1997 (2007).

16. F. J. Owens, *Nanoscale Res. Lett.* 2, 447 (2007).

17. H. S. Kang, *J. Phys. Chem.* B110, 4621 (2006).

18. R. Wu, C. Peng, L. Liu and Y. P. Feng, *Appl. Phys. Lett.* 86, 122510 (2005).

19. M. Miller and F. J. Owens, *Solid State Commun.* 151, 1001 (2011).

20. B. P. C. Rao, et al. *Phil. Mag.* 90, 3463 (2001).

21. Y. Wang, et al. *Nanoletters* 9, 330 (2009).

22. R. R. Nair, et al. *Nat. Phys.* 8, 199 (2012).

23. M. Miller and F. J. Owens, *Chem. Phys. Lett.* 570, 42 (2013).

24. W. Li, et al. *Phys. Chem. Chem. Phys.* 12, 13699 (2012).

25. K. Nakada, M. Fujita, G. Dresselhaus and M. S. Dresselhaus, *Phys. Rev.* B54, 1795 (1996).

26. V. L. Joly, et al. *Phys. Rev.* B81, 245428 (2010).

27. J. Zhou, et al. *Nano Lett.* 9, 3867 (2009).

28. O. V. Yazyev, *Rept. Prog. Phys.* 73, 056501 (2010).

6

NANOSTRUCTURED MAGNETIC SEMICONDUCTORS

6.1 ELECTRON–HOLE JUNCTIONS

Silicon, the major component of most transistor devices, has four valence electrons, which are shared in the covalent bonds with four nearest neighbor silicon atoms in the silicon lattice. If the lattice is doped with a phosphorous atom, which has five valence electrons, an extra electron is available to contribute to the conductivity of the lattice. The donor energy level of this extra electron lies in the band gap just below the conduction band, as shown in Figure 6.1, such that a very small amount of thermal energy, kT, will excite it to the conduction band. This kind of doped semiconductor is called an N-type semiconductor. If silicon atoms are replaced by aluminum, which has three valence electrons, one bond will be missing an electron and is referred to as a hole. The acceptor energy level of the hole, as shown in Figure 6.1, is just above the valence band, and this hole can also contribute to the conductivity of the material. This kind of doping is referred to as P doping.

Now, consider a piece of silicon in which donor impurities (N) are introduced on one side and acceptors (P) on the other side forming what is called a P–N junction as illustrated in Figure 6.2a. When the junction is formed, there is a density gradient of holes and electrons at the interface between the N and P regions. Because of this, holes will diffuse to the N side and electrons to the P side. As a result of the displacement of charge, an electric field will be produced across the junction, and this field will increase until it becomes large enough to prevent further diffusion establishing equilibrium. There will be a net positive charge on the N side and a net negative charge on the P side. As a result, charge cannot move in this region, which is referred to as the depletion region. Now, let us apply a voltage across the junction such that the P side is connected to the negative terminal and the N side to the positive terminal as shown in Figure 6.3. This is referred to as reverse bias. Under this situation, the

Physics of Magnetic Nanostructures, First Edition. Frank J. Owens.
© 2015 John Wiley & Sons, Inc. Published 2015 by John Wiley & Sons, Inc.

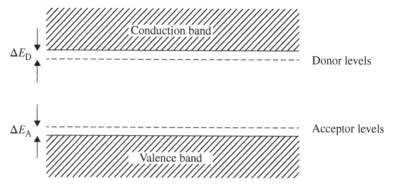

FIGURE 6.1 Schematic of the band gap of a semiconductor showing energy levels for hole and electron doping in the band gap.

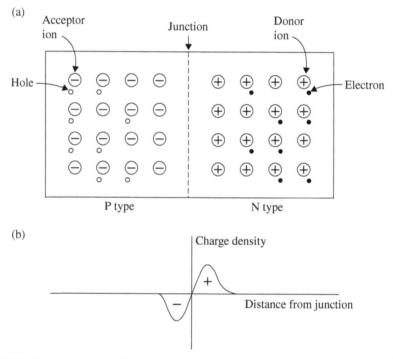

FIGURE 6.2 (a) Schematic illustration of a P–N junction and (b) the charge distribution at the interface due to the net flow of holes and electrons across the junction.

holes and electrons will move away from the P–N interface, and no current can flow across the junction. If we reverse the polarity of the voltage shown in Figure 6.3, current will flow.

Figure 6.4 shows a typical current–voltage relationship for a P–N junction. We see from the figure that the application of +0.15 V produces a current flow in the forward

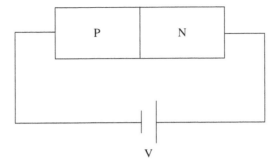

FIGURE 6.3   A P–N junction with a reverse bias voltage applied to it.

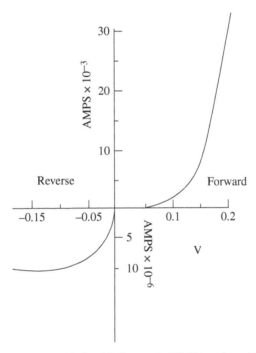

FIGURE 6.4   Current–voltage relationship for a typical P–N junction with an applied forward bias (right) and an applied reverse bias (left). Note the factor of a thousand (10^3) difference between the ordinate scales for the forward and the reverse biases.

direction of 10 ma, which increases very rapidly with a further increase in the applied (bias) voltage. In contrast to this, the application of the reverse bias of −0.15 V produces a negligible flow in the reverse direction of only −0.01 ma, which does not increase with further increases of negative applied voltages. Thus, we can say that, for all practical purposes, a P–N junction does not conduct when subjected to a reverse bias. Such a junction can be used as a rectifier.

6.2 MOSFET

Now, let us consider the basic switching element used on computer chips called a MOSFET, which stands for metal–oxide–semiconductor field-effect transistor. Figure 6.5 shows a simple illustration of a MOSFET, which consists of a P-type semiconductor with an N-type one on both sides. On the top of the P type is a thin layer of silicon oxide that has on top of it a thin layer of a metal connected to an electrical lead to which a voltage can be applied. There is also a metal electrical contact on top of each N-doped semiconductor. If a voltage is applied across the two N types in the figure, no current will flow because one N–P junction is always reversed biased no matter what the polarity of the voltage. Now, if a positive voltage is applied to the metallic contact on the P-doped material called the gate, the holes in the gate will be pushed down to the bottom of the gate, and there will be a narrow channel under the silicon oxide layers, which will be N doped, and thus, current can flow across the junction through this layer when a voltage is applied across the two outer N-doped semiconductors. The device will be in an on state in this case and an off state when no voltage is applied to the gate. Devices like this, which display on and off states representing the 1 and 0 of computer logic, are the fundamental elements of computer chips. Most of the power consumption in a MOSFET occurs when it changes from on to off. In order to reduce power consumption, MOSFETs are arranged in pairs called CMOSs, that is, complementary MOSFETs. The pairs consist of PNP and NPN MOSFETs where the outer layers are the source and the drain. If a negative voltage is applied to the gate of both devices, the PNP switches on and the NPN device switches off. This results in little or no net power dissipation for the complementary pair on switching. Paired MOSFETs or CMOSs are widely used on computer chips. A computer chip is a large array of such switches electrically connected to each other. Nanotechnology can play an important role in increasing the speed of computer chips. The smaller one can make these switches, the MOSFETs, the more switches that can be put on a chip, the closer together they will be, and thus the faster the processing time of the chip. At the present time,

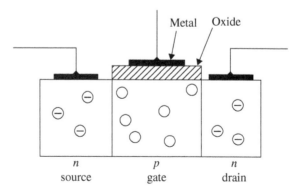

FIGURE 6.5 A schematic illustration of a metal–oxide–semiconductor field-effect transistor called a MOSFET.

MOSFETs are approaching nanometer dimensions. The clock speed of chips has increased by about 29% a year since 1980 because methods have been developed to fabricate smaller and smaller switches on a chip.

6.3 NANOSIZED MOSFETs

At present, the switches are less than a 100 nm long. But how much smaller can they get? It is possible that there may be some small size at which the MOSFET stops working in the desired way. Let us examine some of the challenges in reducing the size of MOSFETs even further. One of the limiting factors in reducing the size is the length of the depletion region. As the size of the device decreases, the width of the depletion region does not change, and hence, the depletion region will become a larger fraction of the gate region. If the depletion region stretches across the gate region, the device will not work because there will be no holes that can be displaced by the gate voltage. Typically, the depletion region in a P–N junction is in the order of 10^{-6} m, a micrometer. Most estimates of the minimum gate length at which a MOSFET will work are about 1/5 of a micrometer, which will allow in excess of 10^9 switches on a chip enabling a clock speed in the gigahertz region. So at present, the fastest desktop computers are getting close to this limit since they have a clock speed close to 4 GHz. To go beyond this new technology may be needed such as molecular-based devices.

However, there are other factors that may limit significant enhancement of speed using MOSFET-like switches such as the interconnects or wires that connect all the on–off switches in the array. The array of switches is actually an RC circuit because the interconnects have resistance, R, and the switches have capacitance, C. In the depletion region at the interface of the P–N junction for a reverse bias, one side of the junction has a positive charge and the other side a negative charge as illustrated in Figure 6.3b. So the MOSFET is effectively a capacitor. The current in an RC circuit has the form

$$I = I_o \exp\left[\frac{-t}{RC}\right] \tag{6.1}$$

The larger the RC, the more quickly the current decays, and it may not reach a neighboring switch. The scaling of the resistance with reduced size of the interconnects contributes to increasing RC. The resistance of a wire of length L and cross-sectional area A is given by

$$R = \frac{\rho L}{A} \tag{6.2}$$

where ρ is the resistivity of the material of which the wire is made and is an intrinsic property of the material. If the thickness, width, and length are reduced by $\acute{a}$, the overall resistance will increase by $\acute{a}$. Hence, reduction in the sizes of the switches and interconnects decreases the decay time of the current flowing in the interconnects. Some possible solutions to the interconnect problem are to use metal carbon nanotubes as the connecting wires because they have very low resistance and angstrom diameters. Graphene nanoribbons could also be used.

Another problem that occurs when MOSFETs are reduced to nanometer dimension is leakage. The thin oxide layer under the metal gate shown in Figure 6.5 is SiO_2, which forms an insulating layer between the metal electrode and the conducting channel in the P-doped silicon of the gate. As chips are packed more densely and the size of the MOSFETs reduced to nanometer dimension, the thickness of the SiO_2 layer is reduced. It may be necessary to make the thickness 2 nm or less. When the insulating layer gets this thin, current can tunnel through the layer from the metal to the P-doped region of the gate. This is referred to as leakage current. Below about a 4 nm thickness, the leakage current increases exponentially for approximately every 0.3 nm reduction in the thickness of the insulating layer. This gate current can cause an increase in power consumption of the MOSFET and detrimentally affect its performance. Once leakage current occurs, it can produce defects in the SiO_2 such as microscopic electron traps. When the defect density becomes sufficiently high, the oxide no longer acts as a good insulator and the MOSFET does not function properly. Thus, leakage current reduces the performance lifetime of a MOSFET. There has been some progress in dealing with this problem. A hafnium compound having a high dielectric constant has been found, which may replace SiO_2 and allow nanometer-thick insulating layers.

6.4 DILUTE MAGNETIC SEMICONDUCTORS

As we have discussed previously, the switching elements in the logic circuits of computer processors are semiconducting materials involving an on–off flow of electrons or holes. On the other hand, the storage of information employs hard disks, which depend on the alignment of magnetic nanoparticles. The possibility of combining both the switching and storage of information in one material has generated much interest in the development of semiconductors, which are ferromagnetic referred to as dilute magnetic semiconductors (DMS). Such materials have the potential to increase computer speed. The effort has been primarily focused on semiconducting materials such as GaN, GaP, ZnO, and ZnS doped with ions having unpaired electrons such as Mn^{2+} or Cu^{2+} at levels of 2–6% by weight. A driving force behind the experimental efforts was a theoretical prediction that P-type semiconductors when doped with magnetic ions could be ferromagnetic. It was predicted that the Curie temperature depended on the hole concentration, and at sufficiently high hole concentrations, such doped materials could be ferromagnetic above room temperature. A schematic illustration of the prediction of the dependence of the Curie temperature on hole concentration is shown in Figure 6.6. An essential requirement for a ferromagnetic semiconductor to be useful is that the Curie temperature be well above room temperature.

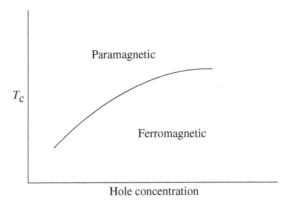

FIGURE 6.6   A schematic plot of the dependence of the Curie temperature of a dilute magnetic semiconductor on the hole concentration.

Ferromagnetism having a T_c of 110 K was observed in GaAs doped with manganese. Later, GaP doped with manganese by ion implantation was reported having an estimated T_c of 385 K. In the ion implantation method, GaP is subjected to a 250 keV beam of Mn ions. This method of synthesis is not inexpensive and not easily scaled up. It was subsequently shown that Mn^{2+} could be doped into ZnO and GaP by a relatively simple sintering process. This process is a solid-state reaction process where MnO and GaP are mixed in the ratio of typically 0.03 of the molecular weight of MnO to one molecular weight of GaP. The material is finely ground for many hours in a mortar and pestle and then pressed into a pellet. The pellet was heated in an aluminum oxide boat in an oven at 500°C for 4 h and then rapidly quenched to room temperature. Since Ga is 3+ ion and Mn is 2+ ion, doping with manganese should hole-dope the material. Evidence for hole doping can be obtained from Raman spectroscopy. Figure 6.7 shows the Raman spectrum of the transverse optical mode (lower-frequency mode) and the longitudinal optic mode (higher-frequency mode) in undoped GaP and Mn-doped GaP [1]. The smaller Raman line is in the doped material. The frequency of the longitudinal optic mode has shifted down in the doped material. This is because the longitudinal optic mode is coupled to the plasma mode. The plasma mode is a collective vibration of the conduction electrons with its frequency proportional to the concentration of conduction electrons. So if the electron concentration is reduced due to hole doping, the plasma frequency is reduced and the longitudinal optic mode frequency decreases. Figure 6.8 shows a measurement of the magnetization of the sample as a function of dc magnetic field at 300 K clearly showing that the material is ferromagnetic at room temperature. Ferromagnetic resonance measurements were used to study the ferromagnetism above room temperature, and the Curie temperature was estimated to be 600 K.

It is to be recalled that the dominant interaction that produces ferromagnetic ordering is the exchange interaction, which is a short-range nearest neighbor interaction between the spins of the paramagnetic ions. Thus, the question arises: how are the spins of the manganese ions coupled to form a ferromagnetic material if they are

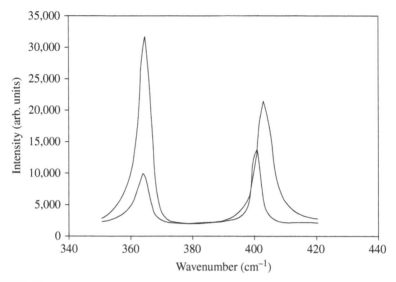

FIGURE 6.7 Raman spectra of the transverse optical (lower frequency) and longitudinal vibrational mode (higher frequency) of undoped GaP (larger spectrum) and Mn^{2+}-doped GaP (smaller spectrum) (Adapted from Ref. [1]).

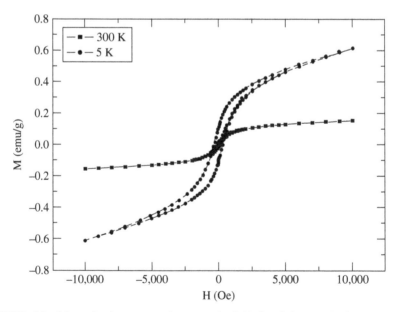

FIGURE 6.8 Magnetization versus dc magnetic field for GaP doped with manganese (Adapted from Ref. [2]).

dispersed in the lattice at low concentrations and widely separated? The explanation of ferromagnetism in DMS is still a matter of research. The present idea is that the conduction holes mediate the coupling between the spins of the paramagnetic ions. The holes move through the lattice acting as go-betweens coupling the spins of the spatially separated paramagnetic ions. The fact that the Curie temperature depends on hole concentration as shown schematically in Figure 6.6 supports this idea. Further support for the idea comes from the observation that shining light on the materials, which causes excitations across the band gap and thus changes hole concentrations, affects the magnitude of the magnetization.

6.5 NANOSTRUCTURING IN MAGNETIC SEMICONDUCTORS

It has been pointed out that the concentration of magnetic dopants such as Mn^{2+} and Cu^{2+} that results in ferromagnetism exceeds the solubility of the dopants in the material. This may result in the formation of nanosized magnetic regions in the semiconductor [2]. Zinc sulfide doped with Cu^{2+} is ferromagnetic at room temperature. Figure 6.9 shows a measurement of the magnetization versus applied dc magnetic field at three temperatures [3]. The material was synthesized from a mixture of one molecular weight of ZnS and 0.04 molecular weight of CuS. The powder was ground into a fine powder and then pressed into a pellet, which was then sintered at 500°C in a tubular oven. Figure 6.10 presents a measurement of the temperature dependence of the magnetization in magnetic field cooled (FC) samples and zero field cooled (ZFC) samples. The splitting between the ZFC and FC curves is characteristic of dynamical

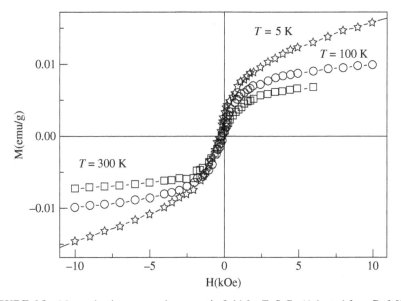

FIGURE 6.9 Magnetization versus dc magnetic field for ZnS:Cu (Adapted from Ref. [3]).

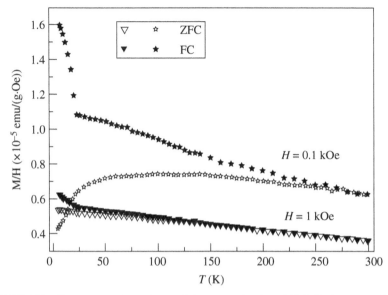

FIGURE 6.10 Temperature dependence of the magnetization of ZnS:Cu for magnetic field and zero field cooled material (Adapted from Ref. [3]).

behavior of magnetic nanoparticles and the existence of a blocking temperature as discussed in Chapter 3 and shown in Figure 3.3 for $CoFe_2O_4$ magnetic nanoparticles. This suggests that the ferromagnetism occurs in nanosized regions of ZnS:Cu. In ferromagnetic materials not having nanosized structure, the irreversibility is due to the presence of domains. The difference of the T dependence of the magnetization above and below 25 K indicates the presence of two contributions to the irreversibility of the magnetization—one where there is long-ranged order and thus domain structure and other from magnetic nanoparticles.

Silicon doped with manganese has been shown to be ferromagnetic at room temperature. This is an important result because silicon is the most widely used material in MOSFETs. There are a number of observations of ferromagnetism in silicon synthesized by different methods such as ion implantation, molecular beam epitaxy, and high-temperature sintering, similar to the process used to make ZnS:Cu [4–7]. This material was studied by magnetic force microscopy (MFM), which is a modification of atomic force microscopy. Figure 6.11 presents a schematic of an atomic force microscope. It consists of a cantilever with a tip on the end whose point is in the order of nanometers. In the case of MFM, the tip is a magnetic material. The surface of the material is moved horizontally below the tip in effect enabling the tip to scan the surface. When the magnetic tip comes to a region that is ferromagnetic, the cantilever undergoes a deflection that is measured by the reflection of a laser beam from the top of the cantilever to a photodiode array. This data is then processed by a computer to form an image of the magnetic structure of the surface of the material. Because the tip is nanosized, magnetic regions of nanometer dimensions can be resolved. Figure 6.12 shows the results of a scan of

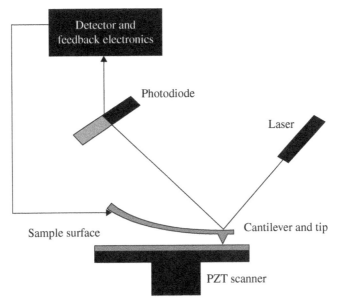

FIGURE 6.11 Schematic illustration of an atomic force microscope or magnetic force microscope if the tip is a magnetic material.

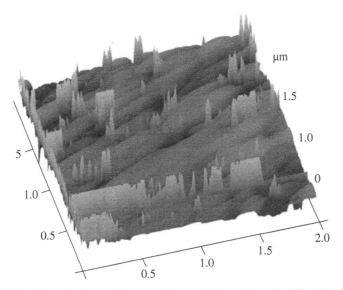

FIGURE 6.12 Magnetic force microscope image of the surface of a Si pellet doped with manganese. The spikes are regions where the material is ferromagnetic (Adapted from Ref. [5]).

the surface of a sintered pellet of silicon doped with manganese [6]. The spikes are regions where the cantilever has undergone deflections and are magnetic. They vary in size but are typically less than 100 nm. This image confirms that the magnetism in DMS is occurring in nanosized regions.

6.6   DMS QUANTUM WELLS

Quantum wells (nanometer-thick) thin films of DMS have been fabricated and shown to be ferromagnetic at room temperature. For example, single crystal films of silicon doped with manganese have been made [4–7]. Figure 6.13 presents a plot of the magnetization versus temperature for this material in a magnetic field of 50 Oe showing the existence of two ferromagnetic phases. There is a low-temperature phase having a Curie temperature of 55 K and a phase having a Curie temperature above room temperature.

6.7   DMS QUANTUM DOTS

Quantum dots were discussed in Chapter 1. Quantum confinement occurs when the dimensions of the nanostructures are in the order of the wavelength of holes or electrons, which carry the current. In a semiconductor, these wavelengths are in the order of a micron, meaning that quantum confinement occurs at a micron or less. For a quantum dot made of a DMS material that is a semiconductor doped with an ion having unpaired electrons such as Mn^{2+}, a number of interesting phenomena emerge. The confinement can enhance the spin–spin exchange interaction. This means that quantum dots may have higher Curie temperatures compared to the bulk material. There have been some studies of quantum dots made of DMS materials such as Mn^{2+}-doped ZnSe and CdS and Co^{2+}- and Ni^{2+}-doped ZnO [8–11]. Figure 6.14 shows a plot of the magnetization versus dc magnetic field for a $ZnO:Co^{2+}$ doped quantum dot showing it is ferromagnetic at room temperature. One

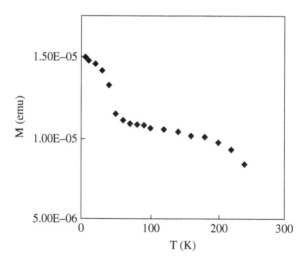

FIGURE 6.13   Magnetization versus temperature in a quantum well of a DMS material of silicon doped with manganese (Adapted from Ref. [5]).

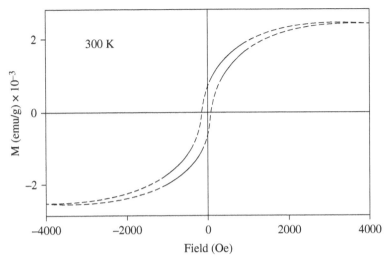

FIGURE 6.14 Plot of the magnetization versus magnetic field at room temperature for a quantum dot of ZnO doped with Co^{2+} (Adapted with permission from Ref. [11]. © 2003 American Chemical Society).

of the interesting observations is that the presence of the magnetic dopant in the dot creates a large internal magnetic field. In the case of Mn^{2+}-doped ZnSe, the field strength was estimated to be 430 T [8]. This internal dc magnetic field has been shown to produce a splitting (Zeeman splitting) of the energy levels of the electron and holes. These are the energy levels just below the conduction band and above the valence band shown in Figure 6.1.

6.8 STORAGE DEVICES BASED ON MAGNETIC SEMICONDUCTORS

DMS can be the basis of magnetic storage devices. One idea that has been proposed is to use them as electrically controlled magnetic switches where a small voltage can order or disorder spins of the paramagnetic ions [12]. The device, illustrated in Figure 6.15, consists of a layer of metal, the gate, on a thin layer of an insulator, which is on top of a DMS followed by a layer of the same material that is not doped and therefore is not ferromagnetic. When a positive voltage is applied to the gate, holes are pushed into the undoped semiconducting layer and the doped layer is not ferromagnetic. When a negative voltage is applied to the gate, the hole concentration increases under the insulating layer and the doped layer becomes ferromagnetic. In effect, a small voltage can control the existence of ferromagnetic ordering in the device. Such a device could be the basis of a magnetic storage device. The fact that the magnetization can also be affected by light opens the possibility of an optical read–write storage device.

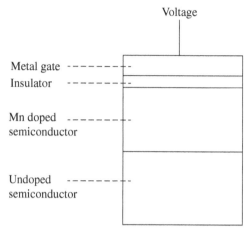

FIGURE 6.15   Illustration of a magnetic storage device employing dilute magnetic semiconductors (Adapted from Ref. [11]).

6.9   THEORETICAL PREDICTIONS OF NANOSTRUCTURED MAGNETIC SEMICONDUCTORS

The field of computational materials science has grown in the last 30 years to become an important part of materials research. Computational materials science refers to theoretically predicting the properties of materials such as the structure and electronic properties. It can and has been used to predict the existence of new materials. For example, the possibility of the existence of C_{60} and its properties such as the vibrational frequencies were predicted long before its discovery. It was also used to predict the possibility that doping semiconductors such as silicon with Mn^{2+} could produce magnetism. Theoretical modeling can also be used to determine the properties of existing materials and is particularly useful in predicting properties that are difficult to determine experimentally. Calculations of elastic constants give results quite close to experimental values and are much easier to obtain compared to experimental methods. The development of computational materials science as an important tool in materials research is largely a result of the development of high-performance computers and theoretical models such as density functional theory (DFT). A critical issue in modeling solids is choice of the size of the solids, meaning the number of atoms or ions in the system. The number of substituents treated must be such that the results give values of properties corresponding to macroscopic systems. From Appendix A, it is seen that a spherical copper nanoparticle having a diameter of 12.54 nm has approximately 10^4 atoms, which is near the limit of numbers that can be handled by present-day computers using the most sophisticated theoretical models. Another limitation is the time scale, which should be less than 10^{-15} s in order to deal with atomic vibrations in the materials.

More recently, theoretical modeling has been applied to nanostructures such as carbon nanotubes and graphene. Perhaps because of the smaller size of nanostructures that typically have less than 10^6 atoms, these limitations may be less of a problem.

A particular objective of this work is to predict new and interesting modifications of nanostructures, which have application potential. This can provide guidance for identifying interesting materials to synthesize.

Here are presented some examples of using computational methods to predict possible magnetic semiconducting nanostructures.

The research to develop magnetic semiconductors has primarily focused on doping existing semiconductors such as silicon or gallium phosphide with Cu or Mn. However, other approaches have been proposed based on predictions from computational modeling. It was seen in Chapter 3 that materials that are antiferromagnetic as macroscopic materials can become ferromagnetic when they are reduced to nanometer dimensions. Thus, reducing an antiferromagnetic semiconductor to nanometer sizes could be an approach to producing a magnetic semiconductor. Semiconductors that have antiferromagnetic phases do exist. One example is copper oxide, which is a narrow band gap P-type semiconductor. It undergoes a transition to an antiferromagnetic state at 230 K. It has been shown that when CuO is reduced to sizes below 10 nm it displays ferromagnetism [13].

Computational modeling has been used to predict that copper chloride could be ferromagnetic when it is reduced to nanometer dimensions. An illustration of the crystal structure is shown in Figure 4.11. The structure consists of parallel chains of copper chloride. As discussed in Chapter 4, the material undergoes a paramagnetic to antiferromagnetic transition at 70 K. The chains are sufficiently far apart such that the magnetic and electronic properties can be calculated assuming isolated $CuCl_2$ chains. Figure 6.16 shows a plot of the calculated highest occupied molecular

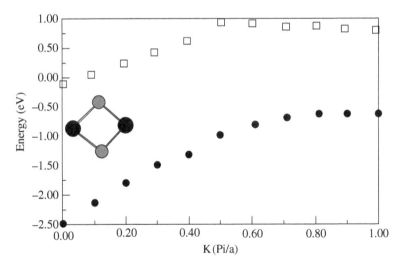

FIGURE 6.16 Calculation of the energy of the highest occupied molecular orbital and the lowest unoccupied molecular orbital of the spin down energy gap versus the K vector for copper chloride. The insert is the unit cell of the chain used to calculate the band gap using periodic boundary conditions. The two copper ions have parallel spins (Adapted from Ref. [14]).

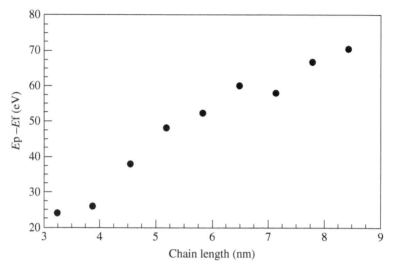

FIGURE 6.17   Plot of the difference in the total calculated energy versus chain length of ferromagnetic chains and paramagnetic chains of $CuCl_2$. E_f is the total energy of the ferromagnetic chains, and E_p is the total energy of the paramagnetic chains. The ferromagnetic chains have lower energy (Adapted from Ref. [14]).

orbital (HOMO) and the lowest unoccupied molecular orbital (LUMO) versus the K vector calculated using periodic boundary conditions and DFT [14]. The separation between the two levels is the band gap. The gap at $K=0$, the center of the zone, is obtained to be 2.5 eV indicating the material is semiconducting. Figure 6.17 shows a plot of the calculated difference in total energy of the paramagnetic state, E_p, and the ferromagnetic state E_f, versus chain length in the nanometer region. These results indicate the ferromagnetic state is the more stable state and that $CuCl_2$ chains of nanometer length could be ferromagnetic. Since macroscopic-sized copper chloride is not ferromagnetic, $E_p - E_f$ should eventually decrease to zero with increasing chain length.

Another example of predicting ferromagnetism in semiconducting materials involves boron nitride nanoribbons. Figure 6.18 shows a plot of the calculated HOMO–LUMO energy gap versus ribbon length of a zigzag boron nitride nanoribbon similar to the graphene shown in Figure 5.17a [15]. The calculation used DFT with the local spin density approximation employing periodic boundary conditions. The results show that at the longer lengths the ribbon becomes semiconducting. Interestingly, the calculation for an armchair ribbon did not show such a decrease with increasing ribbon length. It was shown that when a carbon atom replaced nitrogen, the ribbon was semiconducting and had a larger density of states for the spin down state at the Fermi level. These results suggest that carbon-doped boron nitride nanoribbons could be ferromagnetic semiconductors. Neither of these two results has been experimentally verified; however, the predictions suggest materials to focus synthesis on.

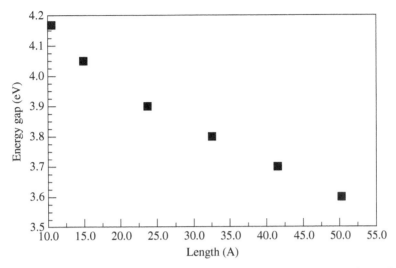

FIGURE 6.18 Calculated band gap of zigzag BN nanoribbons versus ribbon length (Adapted from Ref. [15]).

EXERCISES

6.1 If an ac voltage is applied to the P–N junction shown in Figure 6.4, what would the output current signal look like? Explain your answer.

6.2 What would happen if a negative voltage were applied to the gate of a MOSFET?

6.3 How would the frequency of the longitudinal mode of GaP be affected if doped with a 3+ ion and a 4+ ion?

6.4 The resistivity of copper is 1.725×10^{-8} Ωm. What is the resistance of a cylindrical copper wire having a diameter of 5 nm and a length of 100 nm?

6.5 Aluminum has a resistivity of 2.773×10^{-8} Ωm. Which would make a better interconnect for CMOS transistors, aluminum or copper? Explain the answer.

6.6 Experiments indicate that the magnetism in DMS materials is occurring in nanosized regions. How would the size of these regions affect the magnetization and the coercivity?

REFERENCES

1. F. J. Owens, *J. Phys. Chem. Solid.* 66, 973 (2005).
2. T. Dietl, *J. Appl. Phys.* 103, 07D111 (2008).
3. F. J. Owens, et al. *J. Phys. Chem. Solid.* 72, 648 (2011).
4. M. Boldue, et al. *Phys. Rev.* B71, 033305 (2005).

5. (a)F. M. Zhang, et al. *Appl. Phys. Lett.* 85, 786 (2004);(b)F. M. Zhang et al. *J. Mag. Mag. Mater.* 22, 201 (2008).

6. R. Patel and F. J. Owens, *Solid State Commun.* 152, 603 (2012).

7. S. H. Chiu, H. S. Hsu and J. C. A. Huang, *J. Appl. Phys.* 103, 07D110 (2008).

8. D. J. Norris, N. Yao, F. T. Charnock and T. A. Kennedy, *Nano Lett.* 1, 3 (2001).

9. D. M. Hoffman, et al. *Solid State Commun.* 114, 547 (2000).

10. K. M. Hanif, R. W. Meulenberg and G. F. Strouse, *J. Am. Chem. Soc.* 124, 11495 (2002).

11. D. A. Schwartz, et al. *J. Am. Chem. Soc.* 125, 13205 (2003).

12. H. Ohono, et al. *Nature*, 408, 944 (2000).

13. A. Punnoose, H. Magnone and M. S. Seehra, *Phys. Rev.* B64, 174420 (2001).

14. F. J. Owens, J. Nanoparticles, 2013, 758473.

15. A. J. Du, S. C. Smith and G. Q. Lu, *Chem. Phys. Lett.* 447, 181 (2007).

7

APPLICATIONS OF MAGNETIC NANOSTRUCTURES

In this chapter, some present and potential nonmedical applications of magnetic nanostructures are discussed.

7.1 FERROFLUIDS

Ferrofluids, also called magnetofluids, are fluids containing typically 10 nm magnetic particles coated with a surfactant to prevent aggregation. Typical fluids used are transformer oil and kerosene. The nanoparticles are single-domain magnets, and in zero magnetic field, at any instant of time, the magnetization vector of each particle is randomly oriented so the liquid has a zero net magnetization. When a dc magnetic field is applied, the magnetizations of the individual nanoparticles all align with the direction of the field, and the fluid acquires a net magnetization. Typically, ferrofluids employ nanoparticles of magnetite, Fe_3O_4. The magnetization curve for a ferrofluid made of 6 nm Fe_3O_4 particles exhibits almost immeasurable hysteresis. Ferrofluids are soft magnetic materials that are superparamagnetic. Interestingly, suspensions of magnetic particles in fluids have been used since the 1940s in magnetic clutches, but the particles were larger, having micron dimensions. Application of a dc magnetic field to this fluid causes the fluid to congeal into a solid mass, and in the magnetic state, the material is not a liquid. A prerequisite for the existence of a ferrofluid is that the magnetic particles have nanometer sizes. Such fluids will not congeal into solids when a dc magnetic field is applied.

Ferrofluids have a number of interesting properties, such as magnetic field-dependent anisotropic optical properties. Analogous properties are observed in liquid crystals, which consist of long molecules having large electric dipole moments, which can be oriented by the application of an electric field in the fluid phase. Electric

Physics of Magnetic Nanostructures, First Edition. Frank J. Owens.
© 2015 John Wiley & Sons, Inc. Published 2015 by John Wiley & Sons, Inc.

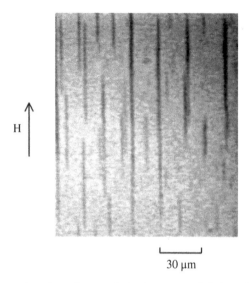

H

30 μm

FIGURE 7.1 Picture taken through an optical microscope of chains of magnetic nanoparticles formed in a film of a ferrofluid when the dc magnetic field is parallel to the plane of the film (With permission from Ref. [1]).

field-modulated birefringence or double refraction of liquid crystals is widely used in optical devices, such as liquid crystal displays in digital watches and screens of portable computers. This suggests a potential application of ferrofluids employing magnetic field-induced birefringence. To observe the behavior, the ferrofluid is sealed in a glass cell having a thickness of several micrometers. When a dc magnetic field is applied parallel to the surface and the film is examined by an optical microscope, it is found that some of the magnetic particles in the fluid agglomerate to form needlelike chains parallel to the direction of the magnetic field. Figure 7.1 shows a picture of the chains taken through an optical microscope. As the magnetic field increases, more particles join the chains, and the chains become thicker and longer. The separation between the chains also decreases. Figure 7.2a and b shows plots of the chain separation and the chain width as a function of the dc magnetic field. When the field is applied perpendicular to the face of the film, the ends of the chains arrange themselves in the pattern shown in Figure 7.3, which is also a picture taken through an optical microscope. Initially, at low fields, the ends of the chains are randomly distributed in the plane of the fluid. As the field increases, a critical field is reached where the chain ends become ordered in a two-dimensional hexagonal array as shown in Figure 7.3.

The formation of the chains in the ferrofluid film when a dc magnetic field is applied makes the fluid optically anisotropic. Light, or more generally electromagnetic, waves have oscillatory magnetic and electric fields perpendicular to the direction of propagation of the beam. Light is linearly polarized when these vibrations are confined to one direction perpendicular to the direction of propagation, rather than having random transverse directions. If the plane contained an xy coordinate system, vibration of the E

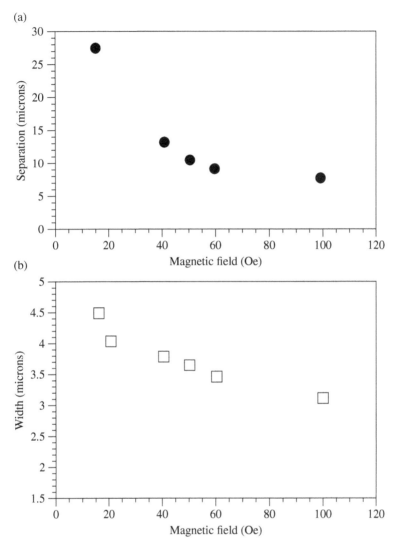

FIGURE 7.2 (a) Plot of separation of magnetic nanoparticle chains versus strength of a magnetic field applied parallel to the surface of the film and (b) plot of the thickness of the chains as a function of dc magnetic field strength. An oersted corresponds to 10^{-4} Tesla (Adapted from Ref. [1]).

vector parallel only to Y would be an example of linearly polarized light. When linearly polarized light is incident on a magnetofluid film to which a dc magnetic field is applied, the light emerging from the other side of the film is elliptically polarized. Elliptically polarized light occurs when the E and H vibrations around the direction of propagation are confined to two mutually perpendicular planes, and the vibrations in each plane are out of phase. This is called the Cotton–Mouton effect. An experimental

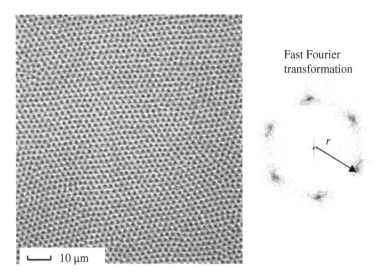

FIGURE 7.3 Optical microscope picture of the ends of chains of magnetic nanoparticles in a ferrofluid film when the dc magnetic field is perpendicular to the surface of the film. The field strength is high enough to form the hexagonal lattice configuration (With permission from Ref. [1]).

arrangement to investigate this effect is shown in Figure 7.4. A linearly polarized He–Ne laser beam is incident on the magnetofluid film. A dc magnetic field is applied parallel to the plane of the film. To examine the polarization of the light emerging from the film, another polarizer called an analyzer is placed between the film and a light detector, which could be photomultiplier or silicon semiconductor. The intensity of the transmitted light is measured as a function of the orientation of the polarizing axis of the analyzer given by the angle η in the figure. Figure 7.5 shows that the transmitted light intensity depends strongly on the angle η. These effects could be the basis of optical switches where the intensity of transmitted light is switched on and off using a dc magnetic field or by changing the orientation of the polarizer.

Ferrofluids can also form magnetic field tunable diffraction gratings. Diffraction is the result of interference of two or more light waves of the same wavelength traveling paths of slightly different lengths before arriving at a detector such as a photographic film. When the path length differs by half a wavelength, the waves destructively interfere, resulting in a dark band on the film. When the path lengths differ by a wavelength, then the waves constructively interfere, producing a bright band on the film. A diffraction grating consists of small slits separated by distances of the order of the wavelength of the incident light. We saw previously that when a dc magnetic field of sufficient strength is applied perpendicular to a magnetofluid film, an equilibrium two-dimensional hexagonal lattice is formed with columns of nanoparticles occupying the lattice sites. This structure can act as a two-dimensional optical diffraction grating that diffracts incoming visible light. Figure 7.6 shows the chromatic (colored) rings of light and darkness resulting from the diffraction and

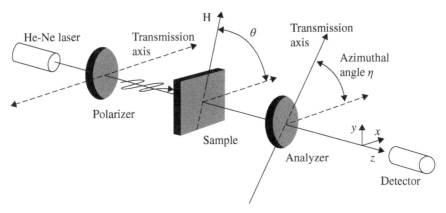

FIGURE 7.4 Experimental arrangement for measuring optical polarization effects in a ferrofluid film that has a dc magnetic field H applied parallel to its surface (With permission from Ref. [1]).

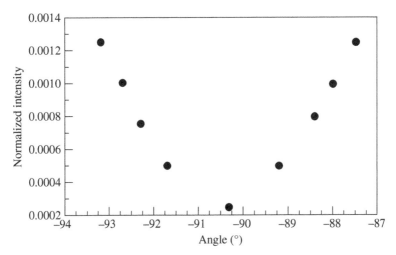

FIGURE 7.5 Intensity of light transmitted through the analyzer in Figure 7.4 versus the azimuthal angle η of the light beam in zero field and in a 50 Oe (0.005 tesla) applied magnetic field H. The rate of increase of the magnetic field is 50 Oe/S (Adapted from Ref. [1]).

interference when a focused parallel beam of white light is passed through a magnetic fluid film that has a magnetic field applied perpendicular to it. The diffraction pattern is determined by the equation

$$d \sin \Theta = n\lambda \tag{7.1}$$

where d is the distance between the chains of nanoparticles, Θ is the angle between the outgoing light and the direction normal to the film, n is an integer, and λ is the wavelength of the light. It was seen earlier that the distance d between the chains

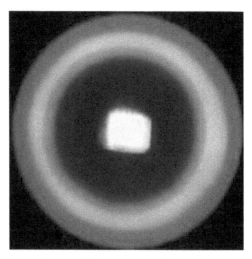

FIGURE 7.6 Chromatic rings resulting from the diffraction and interference of a beam of white light incident perpendicular to a ferrofluid film in a perpendicular dc applied magnetic field. The original figure was in color. The colors of the rings are red, yellow, green, and blue from the outside to the inside (Adapted from Ref. [1]).

depends on the strength of the applied dc magnetic field. In effect, we have a tunable diffraction grating, which can be adjusted to a specific wavelength by changing the strength of the dc magnetic field.

Ferrofluids have a number of present commercial uses. They are employed as contaminant exclusion seals on hard drives of personal computers and vacuum seals for high-speed high-vacuum motorized spindles. In this latter application, the ferrofluid is used to seal the gap between the rotating shaft and the pole piece support structure. The seal consists of a few drops of ferrofluid in the gap between the shaft and a cylindrical permanent magnet that forms a collar around it. The fluid forms an impermeable O-ring around the shaft while allowing rotation of the shaft without significant friction. Seals of this kind have been utilized in a variety of applications. Ferrofluids are used in audio speakers in the voice gap of the driver to dampen moving masses.

7.2   MAGNETIC STORAGE (HARD DRIVES)

The study of magnetic materials, particularly of films made of nanomagnets, is driven by the desire to increase storage space on magnetic storage devices such as hard drives in computers. The basic information storage mechanism involves alignment of the magnetization in one direction of a very small region on the magnetic tape called a bit. To achieve a storage of 10 gigabits (10^{10} bits) per square inch, a single bit would be approximately 1 μm long and 70 nm wide. The film thickness could be about 30 nm. Existing magnetic storage devices such as hard drives are

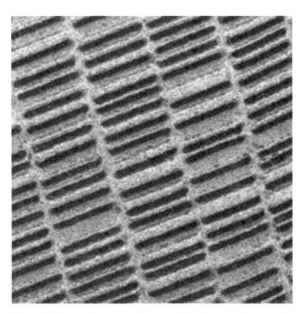

FIGURE 7.7 Magnetic force microscope image of the elongated magnetic nanoparticles on a hard drive layer.

based on single layers of nanosized crystals of cobalt chromium alloys such as CoPtCrTa. Figure 7.7 shows a magnetic force microscope image of a typical hard drive layer.

There is an intense research effort to increase the storage capacity of hard drives. In order to achieve this, smaller stable nanoparticles having high coercivity, low magnetization, and negligible magnetic coupling between the particles are required. The interaction between the particles can be reduced by coating them with a nonmagnetic material. The effort to reduce the particle size of the CoPtCrTa alloys in order to increase storage capability has resulted in particle sizes down to 8–10 nm. This is close to the size where the particles become superparamagnetic. At this size, the thermal energy at room temperature, kT, is close to the magnetic anisotropic energy, which means the particles can undergo thermal fluctuation and become disordered resulting in loss of stored information. The present effort to increase storage capacity is focusing on identifying smaller particles that have higher magnetic anisotropic energies. One candidate material is FePt. It has a face-centered tetragonal crystal structure illustrated in Figure 7.8. The structure consists of two-dimensional layers of iron atoms (arrows in figure) between layers of Pt atoms. The easy direction of magnetic field-induced alignment of the magnetic moments is parallel to the tetragonal C axis. FePt had a very large anisotropic constant, K, which is about 50–100 times larger than the CoPtCrTa alloys. This means much smaller particles in the order of 3.0 nm can be used on the layers in the hard drive.

As discussed previously, when nanoparticles reach a certain small diameter, the magnetic vectors of the individual atoms of the particle become aligned in the ordered

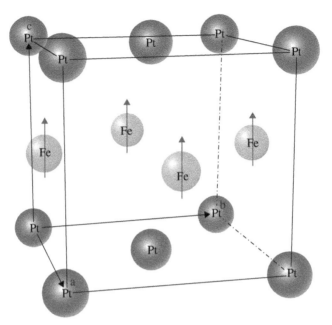

FIGURE 7.8 Tetragonal unit cell of FePt. Arrows on the iron atoms indicate direction of magnetic moments in the ferromagnetic phase.

pattern of a single domain in the presence of a dc magnetic field, eliminating the complication of domain walls and regions having the magnetization in different directions. The Stoner–Wohlfarth (SW) model has been used to account for the dynamical behavior of small nanosized elongated magnetic grains. Elongated grains are generally the type used in magnetic storage devices. The SW model postulates that in the absence of a dc magnetic field ellipsoidal magnetic particles can only have two stable orientations for their magnetization, either up or down with respect to the long axis of the magnetic particles, as illustrated in Figure 7.9. The energy versus orientation of the vectors is a symmetric double well potential with a barrier between the two orientations. The particle may flip its orientation by thermal activation, due to an Arrhenius process where the probability P for reorientation is given by

$$P \sim \exp\left(\frac{-E}{k_{\mathrm{B}}T}\right) \qquad (7.2)$$

where E is the height of the energy barrier between the two orientations. The particle can also flip its orientation by a much lower probability process called quantum mechanical tunneling. This can occur when the thermal energy $k_{\mathrm{B}}T$ of the particle is much less than the barrier height. This process is a purely quantum mechanical effect resulting from the fact that solution to the wave equation for this system predicts a small probability for the up state of the magnetization to change to the down state.

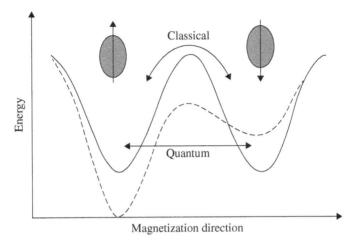

FIGURE 7.9   Sketch of double well potential showing the energy plotted versus the orientation of the magnetization for up and down orientations of elongated magnetic nanoparticles in the absence (——) and presence (----) of an applied magnetic field (Adapted from Ref. [2]).

If a magnetic field is applied, the shape of the potential changes, as shown by the dashed line in Figure 7.9, and one minimum becomes unstable at the coercive field.

The SW model provides a simple explanation for many of the magnetic properties of small magnetic particles, such as the shape of the hysteresis loop. However, the model has some limitations. It overestimates the strength of the coercive field because it only allows one path for reorientation. The model assumes that the magnetic energy of a particle is a function of the collective orientation of the spins of the magnetic atoms in the particle and the effect of the applied dc magnetic field. This implies that the magnetic energy of the particle depends on its volume. However, when particles are in the order of 6 nm in size, most of their atoms are on the surface, which means they can have very different magnetic properties than larger grain particles. It has been shown that treating the surfaces of nanoparticles of α–Fe that are 600 nm long and 100 nm wide with various chemicals can produce variations in the coercive field by as much as 50%, underlining the importance of the surface of nanomagnetic particles in determining the magnetic properties of the grain. Thus, the dynamical behavior of very small magnetic particles is somewhat more complicated than predicted by the SW model and remains a subject of continuing research.

7.3   ELECTRIC FIELD CONTROL OF MAGNETISM

Electric field control of magnetism would open up the possibility of new devices. An electric field cannot directly affect a magnetic moment of an atom. One approach to finding a method to use an electric field to control magnetic properties is to find materials that have both magnetic and ferroelectric properties and in which there is a coupling between the ferroelectric order and the magnetic order. A ferroelectric

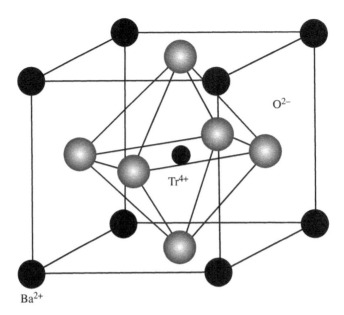

O^{2-}

Tr^{4+}

Ba^{2+}

FIGURE 7.10   Cubic crystal structure of $BaTiO_3$ above the ferroelectric phase.

displays a net dipole moment when no electric field is present. In such a material, the center of positive charge does not coincide with the center of negative charge. Barium titanate is the most thoroughly studied and used ferroelectric. It becomes ferroelectric at 156 K, the Curie temperature. Figure 7.10 shows the cubic crystal structure above 156 K. Below this temperature where the crystal develops a dipole moment, the structure is slightly distorted with the Ba^{2+} and Ti^{4+} displaced parallel to the [001] direction relative to the O^{2-} ions. Figure 7.11 shows a plot of the dipole moment along the [001] direction versus temperature. A plot of the polarization versus electric field resembles the plot of the magnetization versus magnetic field shown in Figure 2.8 and displays a similar hysteresis. There are materials such as $BiFeO_3$ (BFO) that display both magnetic ordering and ferroelectricity. This material becomes ferroelectric at 1103 K and antiferromagnetic at 643 K. In bulk BFO, there is no significant coupling between the magnetic order and the ferroelectric order. Thus, applying an electric field at room temperature does not affect the magnetic order. However, when BFO is fabricated into nanometer-thick epitaxial films, the polarization increases and there is a sizable coupling between the magnetic and ferroelectric orders. This has been exploited to fabricate a device that allows control of the magnetization of the material with an electric field. A 3 nm thick single crystal film of $La_{0.7}Sr_{0.3}MnO_3$ (LSMO) that is ferromagnetic below 374 K was deposited by a pulsed laser technique on the (001) surface of $SrTiO_3$. This was followed by deposition of 200 nm thick epitaxial layer of BFO. The films were patterned into field-effect devices as shown in Figure 7.12a. When a negative voltage is applied to the gate, the ferroelectric polarization in the BFO is reversed, and the magnetization curves of the LSMO are altered as shown schematically in Figure 7.12b.

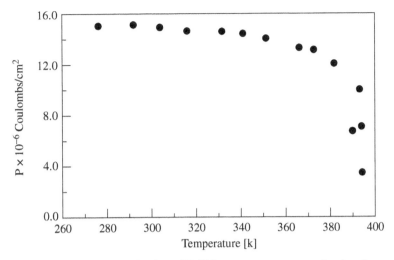

FIGURE 7.11 Plot of the polarization of BaTiO$_3$ versus temperature showing the onset of the ferroelectric phase (Adapted from Ref. [3]).

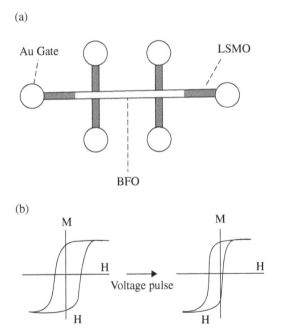

FIGURE 7.12 (a) Illustration of a field-effect device that allows control of magnetic properties of a nanosized ferromagnetic material with a voltage. (b) Illustrates the effect of a voltage pulse on the magnetization curves of LSMO in the device shown in (a) (Adapted from Ref. [4]).

7.4   MAGNETIC PHOTONIC CRYSTALS

A photonic crystal consists of a lattice of dielectric materials with separations on the order of the wavelength of visible light and periodic in the strength of the dielectric constant. Such crystals have interesting optical properties. In photonic crystals, there is an energy gap, meaning that there are certain wavelengths of light that will not propagate in the lattice. This is a result of Bragg reflections. A one-dimensional photonic crystal consists of layers of rectangular-shaped materials of nanometer thickness having alternating large and small dielectric constants. Such a structure is illustrated in Figure 7.13a. The path difference between two waves reflected from adjacent planes is $2d \sin \Theta$, where Θ is the angle of incidence of the wave vector of the light to the planes. If the path difference, $2d \sin\Theta$, is a half wavelength, the reflected waves will destructively interfere and cannot propagate in the lattice resulting in an optical energy gap. This is a result of the lattice periodicity and the wave nature of the light.

For visible light, this requires a lattice dimension of about 500 nm. Thus, photonic crystals that function in the range of visible light are nanostructures. Such crystals have to be artificially fabricated by methods such as electron beam lithography or x-ray lithography, which are discussed in Chapter 9. Photonic crystals have many applications. For example, a crystal with a band gap could be a filter for certain wavelengths, passing only those frequencies not in the gap.

A one-dimensional magnetic photonic crystal consists of nanometer layers of materials having alternating high and low dielectric constants, one of which is a magnetic material. Figure 7.13b illustrates a photonic crystal having a single magnetic layer incorporated into it. The presence of the magnetic layer allows further control of the properties of the light passing through the photonic crystal. Magnetic materials have two different influences on light, the Kerr effect and Faraday rotation. The Kerr effect involves the influence of a magnetic material on light reflected from its surface. The reflected light has different polarization properties than the incident light. The changes in the polarization depend on the orientation of the magnetization with respect to the surface of the magnetic material. The Faraday effect is concerned with light transmitted through a magnetic material. If the incident light is linearly polarized in a given direction, the light emerging from the material will have its plane of polarization rotated. The magnitude of the rotation depends linearly on the component of an applied dc magnetic field parallel to the direction of the propagation of the light. Figure 7.14 illustrates the idea. Thus, introducing magnetic layers into photonic crystals, as illustrated in Figure 7.13b, besides allowing control of the wavelength of the light transmitted allows further control of the nature of the polarization of the light. The effects on polarization can be turned off by raising the temperature above the Curie temperature of the magnetic layer. Photonic crystals with magnetic layers have been fabricated and show large Kerr and Faraday effects. For example, a structure has been fabricated consisting of 10 layers of 110 nm thick SiO_2 (dielectric constant 4.6) and 10 layers of 84 nm thick TiO_2 (dielectric constant 86) on each side of a 40 nm thick layer of a cobalt ferrite [5]. This magnetic photonic crystal displayed a large Faraday rotation of 8.6 °/ μm at 620 nm.

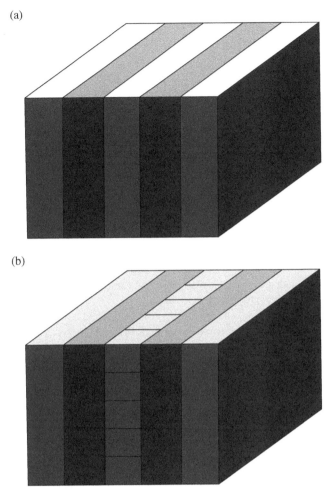

FIGURE 7.13 (a) Illustration of the structure of a one-dimensional photonic crystal, where the dielectric constant of the layers alternates in magnitude. (b) Illustration of a magnetic photonic crystal where the center layer (cross hatched) is a ferromagnetic material.

7.5 MAGNETIC NANOPARTICLES AS CATALYSTS

Catalysis involves the modification of the rate of a chemical reaction, usually a speeding up or acceleration of the reaction rate, by the addition of a substance called a catalyst that is not consumed during the reaction. Ordinarily, the catalyst participates in the reaction by combining with one or more of the reactants, and at the end of the process, it is regenerated without change. In other words, the catalyst is being constantly recycled as the reaction progresses. There are two main types of catalysts. Homogeneous catalysts are dispersed in the same phase as the reactants, the dispersal ordinarily being in a gas or a liquid solution. Heterogeneous catalysts are in a

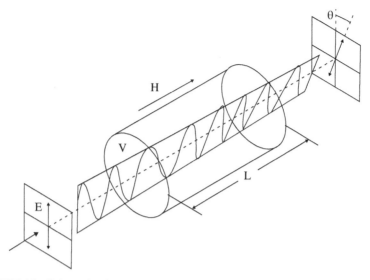

FIGURE 7.14 Schematic of the Faraday effect that causes a rotation of the plane of polarized light transmitted through a magnetic material.

different phase than the reactants, separated from them by a phase boundary. Heterogeneous catalytic reactions usually take place on the surface of a solid catalyst, such as silica or alumina, which have very high surface areas that typically arises from their porous or spongelike structure. These catalysts have their surfaces impregnated with acid sites, or coated with a catalytically active material such as platinum, and the rate of the reaction tends to be proportional to the accessible area of a platinum coated surface.

As discussed earlier, nanoparticles have a large fraction of their atoms on the surface. The chemical activity of a heterogeneous particle catalyst is proportional to the ratio of the surface area to the volume, A/V. Since nanoparticles have a high A/V ratio, they have the potential to be good catalysts. However, magnetic nanoparticles that can also be catalysts can be an exception to this rule. Rhodium, which as discussed in Chapter 3 is paramagnetic in the bulk, becomes ferromagnetic when it has nanometer dimensions. Figure 7.15 shows the effect of rhodium nanoparticles on the dissociation of carbon monoxide. Rhodium nanoparticles of various sizes, characterized by the number of Rh atoms per particle, were deposited on alumina (Al_2O_3) films. The rhodium was given a saturation carbon monoxide (CO) coverage. Then the material was heated from 90 to 500 K (circles) or from 300 to 500 K (squares), and the amount of atomic carbon formed on the rhodium measured indicating dissociation rate for each particle size. While generally the rate of reaction scales with the surface area, it is interesting to note that the catalytic activity in the nanometer range does not scale with the number of atoms on the surface.

Further research is needed to determine whether this is due to the magnetism of the particles.

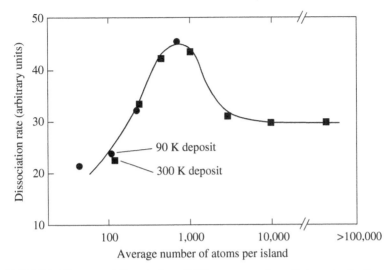

FIGURE 7.15 The rate of dissociation of CO versus number of atoms in rhodium nanoparticles showing the catalytic effect of a magnetic nanoparticle.

7.6 MAGNETIC NANOPARTICLE LABELING OF HAZARDOUS MATERIALS

The detection of hazardous materials such as explosives at airports or at the entrances of secure facilities is a difficult problem. A number of ideas have been considered such as using Raman spectroscopy to identify the chemical or doping the chemical with a luminescent material. The drawback in these two approaches is that they require the detecting light to be directly incident on the chemical. Generally, the explosive will be contained in a package or suitcase. Because magnetic particles can easily be detected even inside packages or suitcases, another approach that has been investigated is to incorporate magnetic particles into explosive materials. There has been some preliminary research on this approach. Initial efforts focused on trinitrotoluene (TNT), a commonly used military explosive. One of the advantages of TNT is it has a low melting point of 81°C, which allows the particles to be incorporated in the molten phase. Ten nanometer particles of Fe_3O_4 were incorporated into the TNT melt. The melt was then subjected to sonication to disperse the particles. The molten TNT was then allowed to cool to room temperature in the presence of a 4000 Gauss dc magnetic field. This aligns the nanoparticles such that the direction of maximum magnetization is parallel to the direction of the applied magnet field. The material was studied by FMR and AC susceptibility. Figure 7.16 shows the FMR spectrum of the Fe_3O_4 in TNT for the magnetic field parallel and perpendicular to the direction of the cooling magnetic field. Figure 7.17 presents a plot of the magnetization versus dc magnetic field normalized to its value at 3450 Gauss. These results show that magnetic nanoparticles can readily be detected in explosives by FMR and AC susceptibility

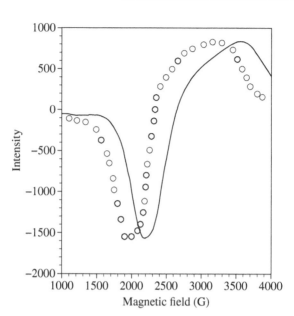

FIGURE 7.16 FMR spectrum of 10 nm Fe_3O_4 nanoparticles in TNT for the dc magnetic field parallel (○) and perpendicular (−) to the direction of the field in which the material was cooled below the melt. From F. J. Owens, unpublished observations.

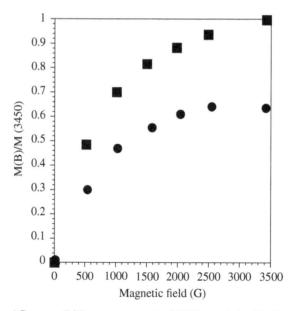

FIGURE 7.17 AC susceptibility measurement of TNT containing Fe_2O_3 nanoparticles for the dc magnetic field parallel (■) and perpendicular (●) to the cooling magnetic field. From F. J. Owens, unpublished observations.

measurements. Different kinds of magnetic nanoparticles having different magnetic properties could be incorporated into different explosives to enable identification of the kind of material.

EXERCISES

7.1 If a magnetic field of 100 G were applied perpendicular a thin film of a ferrofluid and a beam of white light was incident on the film and perpendicular to the film, what wavelength of light would be diffracted at 20° from the normal?

7.2 Why does coating magnetic nanoparticles with a nonmagnetic material reduce the magnetic interaction between the particles?

7.3 Provide an example of an application of the electric field control of magnetism.

7.4 What wavelength of light would not propagate in a one-dimensional photonic crystal having a lattice parameter of 400 nm? Explain the reason for the answer.

7.5 In a magnetic photonic crystal where the magnetic material is Gd, at what temperature would no Kerr effect be observed? Why?

REFERENCES

1. H. E. Hornig, et al. *J. Phys. Chem. Solid.* 62, 1749 (2001).
2. D. D. Awschalom and D. P. DiVincenzo, *Phys. Today* (56) 44 (1995).
3. W. J. Merz, *Phys. Rev.* 76, 1221 (1949).
4. J. M. Wu, et al. *Phys. Rev. Lett.* 110, 067202 (2013).
5. E. Takeda, et al. *J. Appl. Phys.* 87, 6782 (2000).

8

MEDICAL APPLICATIONS OF MAGNETIC NANOSTRUCTURES

8.1 TARGETED DRUG DELIVERY

Cancer encompasses a number of different diseases in which cells grow in an unregulated manner. The cells divide and increase in size in an uncontrolled way producing localized areas of growth called tumors. Cancer can spread to other parts of the body via the lymphatic system or the bloodstream. Noncancerous tumors called benign tumors do not grow uncontrollably and cannot spread throughout the body. The most common treatment for cancer is the use of chemotherapeutic drugs (cytotoxic chemicals). These kill the cancer cells but unfortunately also kill healthy cells, thus producing unpleasant side effects.

Magnetic nanoparticles can provide a way to target drug delivery to the tumor eliminating the negative effects on healthy cells. Tumors need nourishment such as oxygen in order to grow. To do this, the tumors rapidly grow new blood vessels. This is called angiogenesis. Because of their rapid growth, the vessels have more defects than normal vessels in the form of holes on the walls. Typically, these holes have sizes ranging from a few hundred nanometers to microns. The holes in normal blood cells are much smaller ranging from 2 to 6 nm. This means that incorporating about 300 nm particles into the bloodstream would allow them to pass into the vessels of tumors but not into normal vessels. In fact, this has been experimentally demonstrated. Thus, by binding nanoparticles with chemotherapy drugs, they could be delivered to the malignant tumor only, avoiding damage to healthy cells. The use of nonmagnetic nanoparticles loaded with chemotherapeutic drugs is presently under development and in some instances already commercially available. An example is Abraxane, a 150 nm particle of albumin, a blood protein surrounding the drug paclitaxel. The nanoparticles are generally designed to release the drug once inside the tumor by application of external stimuli such as heat, ultrasound, or a magnetic field.

Physics of Magnetic Nanostructures, First Edition. Frank J. Owens.
© 2015 John Wiley & Sons, Inc. Published 2015 by John Wiley & Sons, Inc.

Proteins are large biological molecules consisting of chains of amino acid molecules having the general formula $H_2CHRCOOH$, where R is an organic side chain. For example, the amino acid lysine has a side chain $CH_2–CH_2–CH_2–CH_2–NH_3^+$.

Using magnetic nanoparticles to deliver the drugs to the tumor allows control of the flow of the particles in the bloodstream using an external dc magnetic field. Also, once in the tumor, the particles increase the contrast in the magnetic resonance image compared to the surrounding tissue. This will be discussed in the next section. If a dc magnetic field is applied to a collection of magnetic nanoparticles in a fluid such as blood, the particles will not undergo translational motion because there is no net force on them. Instead, the particles will rotate until the direction of maximum magnetization is parallel to the direction of the applied dc magnetic field. The details of this process were discussed in Chapter 7 in the section on ferrofluids. In order to apply a force to the particles, an inhomogeneous magnetic field must be applied. An example of such a magnetic field would be one that varies in strength in some direction. A particle having a magnetic moment μ_z in the z direction in a field that changes in the z direction would have a force on it given by

$$F = \frac{\mu_z dB}{dz} \tag{8.1}$$

Application of such a field can be used to guide the particles to the tumor.

Normally, in an electromagnet, the pole pieces are flat and parallel to each other. Such a magnet will have a homogeneous dc magnetic field perpendicular to the poles provided the poles are not too far apart. An electromagnet that produces an inhomogeneous field can be designed by changing the shape of the poles such as making them in the form of a wedge as shown in Figure 8.1.

8.2 MAGNETIC HYPERTHERMIA

It has been found that if the temperature of a tumor can be raised to about 45°C and held at this temperature for more than 30 min, the tumor can be destroyed. The heat cannot be applied externally to the body surface in the vicinity of the tumor because it would harm healthy cells between the tumor and the surface. Incorporating magnetic nanoparticles into the tumor by direct injection or by methods of targeted drug delivery described earlier can provide a means to heat the tumor. The heating is accomplished by applying a time-varying magnetic field of sufficient strength and frequency. This ac magnetic field would be applied externally to the body. Initially, it was thought that the magnetic particles should not be superparamagnetic but rather display hysteresis. The oscillation of the domains in the particles induced by the AC field produces frictional heating of the particles, which is proportional to the area under the hysteresis curve. The amount of heat generated per unit volume, ΔQ, is equal to the frequency, f, times the area of the hysteresis loop:

$$\Delta Q = \mu_0 f \int H dM \tag{8.2}$$

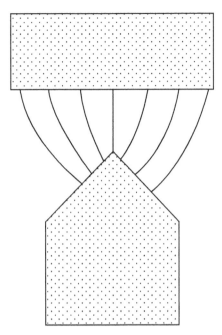

FIGURE 8.1   A permanent magnetic having a wedge-shaped pole that produces an inhomogeneous dc magnetic field.

While most of the research on the use of nonsuperparamagnetic nanoparticles has employed Fe_2O_3 and Fe_3O_4 particles, increased heating could be obtained by use of particles having larger hysteresis such as particles of Nd–Fe–B and Sm–Co. Research on heating of tumors in animals using nonsuperparamagnetic particles has shown that large AC H fields are needed, which could damage healthy human tissue. Thus, more research is needed before this method becomes useful.

An alternative approach has been considered using superparamagnetic nanoparticles. The method uses ferrofluids, suspensions of magnetic nanoparticles in fluids such as hydrocarbons. As discussed in Chapter 7 when a dc magnetic field is applied to a ferrofluid, the particles align with the magnetic field such that the direction of maximum magnetization is parallel to the applied field. When the magnetic field is removed, the magnetization of the individual particles becomes randomly oriented, and the fluid has no net magnetization. The reorientation of the particle magnetization can be characterized by a relaxation time, Ω, which depends on the viscosity, η, of the liquid and the volume of the particle V' given by [1]

$$\Omega = \frac{3\eta V'}{kT} \tag{8.3}$$

The volume of the particle in the liquid, V', is greater than the actual volume of the particle because of the surfactant layer about the particle.

When an AC magnetic field is applied, the magnetic particles undergo oscillatory rotational motion. It turns out that the oscillation of the magnetic moments is not in phase with the AC magnetic field. For small amplitudes and neglecting the interaction between the particles, the response of the magnetization of the ferrofluid to an AC magnetic field can be described by a complex susceptibility, $X = x' + ix''$. The out-of-phase component of the susceptibility, x'', produces heat, which has been shown to be given by [2]

$$Q = \mu_0 \pi f x'' H^2 \tag{8.4}$$

where f is the frequency and H is the amplitude of the ac field. The linear dependence of the heat generated on frequency and the square dependence on H have been experimentally verified [3, 4]. The advantage of using superparamagnetic nanoparticles for generating heat in tumors is that higher heating is produced by lower-amplitude AC magnetic fields. Thus, ferrofluids have the potential to be used to destroy tumors by magnetic heating.

8.3   MAGNETIC SEPARATION

Ferrofluids can be used to both detect the presence of tumor cells in the blood and remove them [5]. The presence of tumor cells in the blood at certain levels is an indication of the possibility of cancer spreading called metastasis. The cells in the blood can be loaded with magnetic nanoparticles by inducing collisions between the cells and the nanoparticles. This can be accomplished by injecting a ferrofluid into the bloodstream and applying a force to the particles using an external magnetic field having a gradient. The presence of the magnetic labeled tumor cells can be detected by passing the cells through a column surrounded by magnets as illustrated in Figure 8.2. As the blood flows through the column, the cells containing the magnetic particles are attracted to the

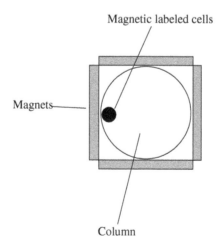

FIGURE 8.2   Illustration of a device that allows separation of magnetically labeled cells from the bloodstream.

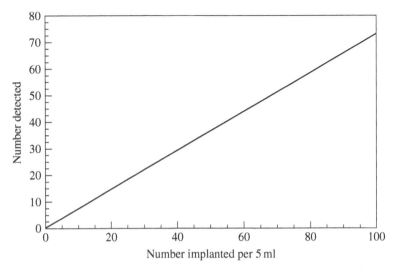

FIGURE 8.3 Plot of the number of cells detected by the device in Figure 8.2 versus the number intentionally added to the bloodstream.

magnets and stick to the walls of the column. This method can be used to detect tumor cells and remove them from the blood. The tumor cells on the walls can be detected by various spectroscopic methods such as atomic force microscopy and magnetometry. Figure 8.3 shows the results of an experiment where fixed amounts of magnetically labeled tumor cells were incorporated into 5 ml of blood and then detected [5]. The results show that approximately 75% of the labeled cells can be detected. The major conclusions resulting from this study, performed on patients, is that tumor cells are present in the blood in the very early stages of cancer and the number of cells detected is a measure of the severity of the cancer and the degree to which it has spread.

8.4 MAGNETIC NANOPARTICLES FOR ENHANCED CONTRAST IN MAGNETIC RESONANCE IMAGING

Nuclear magnetic resonance (NMR) is analogous to electron paramagnetic resonance (EPR) discussed in Chapter 2. Nuclei having an odd mass number have a spin angular momentum $Ih/2\pi$ where I is the nuclear spin. Because the nucleus has both spin and charge, it has a magnetic moment μ_N given by

$$\mu_N = \frac{\gamma_N Ih}{2\pi} \tag{8.5}$$

where γ_N is the magnetogyric ratio of the nucleus having the units radians-s^{-1} Gauss^{-1}. If a dc magnetic field, H, is applied in the z direction, the interaction between the nuclear magnetic moment and the dc field is given by

$$E = -\gamma_N \left(\frac{h}{2\pi} \right) HI_z \qquad (8.6)$$

For a hydrogen atom having a spin ½, there are only two allowed components along the z direction labeled by the quantum numbers $m_s = +1/2$ and $-1/2$. The magnetic field splits the energy into two levels whose separation depends linearly on the dc magnetic field strength similar to the EPR experiment illustrated in Figure 2.2. In NMR, the magnetic field is constant and the incident RF frequency is varied until a transition is induced between the two levels. Note that in the EPR experiment, the frequency is constant and the magnetic field is changed. For NMR, this leads to an absorption of energy given by

$$hf = \gamma_N \left(\frac{h}{2\pi} \right) H \qquad (8.7)$$

Figure 8.4 shows an illustration of a simple NMR spectrometer. The sample is located in a holder between the poles of a dc electromagnet generating a constant magnetic field. The RF radiation that induces a spin flip is applied through a coil around the sample from an oscillator. The frequency of this radiation is changed until an absorption of energy occurs, which is detected by another coil around the sample.

As discussed earlier, the NMR absorption frequency depends on the strength of the magnetic field. If the field is arranged to increase gradually across the dimensions of a sample, then the frequency of the proton NMR absorption would correlate with the location of the hydrogen in the sample. The idea of having the magnetic field vary

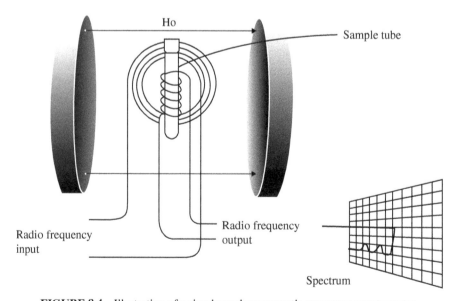

FIGURE 8.4 Illustration of a simple nuclear magnetic resonance spectrometer.

spatially across the sample, thus making a measurement of the frequency in effect a measurement of proton positions in the sample, is the essence of the concept of magnetic resonance imaging (MRI). To illustrate this idea, consider the case of two water drops trapped inside a crystal as shown in Figure 8.5a. The applied magnetic field H is arranged to increase linearly from a value of H_0 at one edge of the sample to a value of H at the other side, as shown in Figure 8.5b, so the increase in the field is directly proportional to the distance x from the left edge of the sample. Since the frequency of energy absorbed is proportional to the strength of the magnetic field, it is also directly proportional to the position across the sample. Thus, a plot of the distribution of trapped water drops across the crystal can be obtained by measuring the intensity of energy absorption at each frequency. For this hypothetical case, Figure 8.5c shows what the spectrum would look like. The small drop produces a

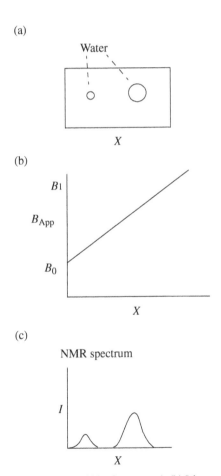

FIGURE 8.5   (a) Two trapped water bubbles in a crystal. (b) Linear variation of dc magnetic field from one side of the crystal to the other. (c) Resulting NMR spectrum from hydrogens of the water droplets.

weak signal and the large drop a strong one. Three-dimensional pictures can be obtained by allowing the magnetic field to vary linearly in all three dimensions.

NMR imaging is a very powerful medical diagnostic technique for a number of reasons. Living tissue is relatively transparent to radio waves, and radio waves do not harm tissue. In addition, living tissue is made of organic molecules that contain hydrogen atoms. The cytoplasm within cells, material between the cells, plasma in the blood vessels, etc., principally consist of water (H_2O), which is the main contributor to the proton signal detected by MRI. MRI is able to distinguish between tissue types, such as kidney and liver tissue, because there is something different about the magnetic resonance spectra in different tissues (if this difference did not exist, then NMR spectra of all tissues would look the same). Fortunately, the viscosity of body fluids, which measures how easily they flow, differs from environment to environment throughout the body. For example, distilled water, which flows easily, has a low viscosity; seawater, as well as water in tissues, contains many salts, and so it has a higher viscosity, while molasses, which flows with considerable difficulty, has a very high viscosity. One of the parameters detected by NMR is called the spin–lattice relaxation time (T_1); this is the time it takes for flipped spins to revert to their original orientations by transferring energy to the lattice vibrations. Another parameter, called the spin–spin relaxation time (T_2), measures the time it takes reversed spins to pass their absorbed energy to nearby spins through mutual spin flips. Both of these relaxation times depend on viscosity; hence, they vary with tissue type. In an NMR experiment, RF energy reverses spin directions, after which they revert to their original orientation, passing absorbed energy to their surroundings, which is called the lattice. The signal is analyzed to provide the two relaxation times and identify characteristics of the tissue being observed. Furthermore, since protons of malignant tumors have unusually long relaxation times, their resonance signals can easily be distinguished from those of normal tissue. As a result, NMR imaging has become a powerful tool for detecting malignant tumors in, for example, the brain. Figure 8.6 shows how well a brain tumor is revealed by MRI. The white area at the bottom of the image is the tumor.

This description of the MRI technique is overly simplified for the purpose of explaining the concept. In the usual MRI technique, the actual measurement is different. Instead of applying continuous radio wave radiation at a single frequency, a sequence of many short pulses within a broad frequency band is applied. The width of the pulse and the separation between pulses depend on the nature of the tissue being examined and the relaxation times of protons in the tissue. Absorbed energy is measured at a fixed time after each pulse and then stored in a computer. This is done repeatedly, with each successive amount added to the previous one. The resulting stored signal, which has many frequency components, is then subjected to a mathematical analysis called Fourier transformation, which identifies how much absorption occurs at each frequency. The result is an image of the distribution of normal and diseased tissues in space.

Reducing T_2 in tissue relative to surrounding areas increases the contrast of area being imaged. T_2 can be reduced by introducing variations, ΔH, in the homogeneity of the applied dc magnetic field; the reduction in T_2 has been shown to be

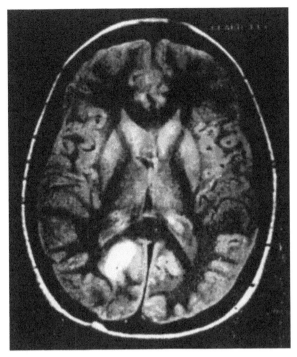

FIGURE 8.6 Magnetic resonance image of a brain containing a tumor. The white region at the bottom is the tumor.

$$\Delta\left(\frac{1}{T_2}\right) = \gamma\Delta H \qquad (8.8)$$

where γ is the gyromagnetic ratio.

The inhomogeneity can be increased by adding magnetic nanoparticles to the tumor or organ being examined. The most commonly used particles are superparamagnetic nanoparticles of iron oxide coated with dextran.

8.5 DETECTION OF BACTERIA

Exposure of people to toxic bacteria such as *E. coli* in produce is a major health concern. There is a need to develop fast, sensitive reliable methods to detect toxic bacteria in food, produce, the environment, and clinical samples. Present methods for detecting bacteria require first growing cultures, which usually takes more than 24 h. A microbiological culture is a medium that enables a specific bacteria to multiply so a sufficient quantity can be obtained for analysis. Not all bacteria can grow in cultures. Use of magnetic nanoparticles has been shown to provide a possible method to rapidly detect some pathogens with high sensitivity.

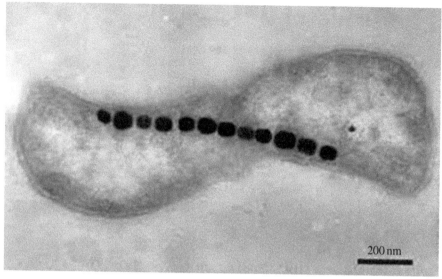

FIGURE 8.7   Electron microscope image of a bacteria containing a chain of ferrite nanoparticles (Reproduced with permission from Dr. Dennis Baylinski).

There are some bacteria referred to as magnetostatic bacteria that have been shown to contain magnetic nanoparticles. Figure 8.7 shows an electron microscope image of bacteria containing a chain of ferrite magnetic nanoparticles. Such bacteria are readily detectable by such methods as SQUID magnetometry, ferromagnetic resonance or NMR. However, most bacteria do not contain magnetic nanoparticles, and a different approach has to be taken to detect them.

The human immune system consists of about 10^{12} cells called lymphocytes, which continuously produce 10^{20} protein molecules called antibodies. These molecules patrol the body to find foreign invaders called antigens. The unique property of the system is the ability of the antibodies to selectively identify and bond to the antigen. The antibody molecule bonds to an antigen when its shape at the edge of the molecule fits some shape of the molecule on the surface of the antigen. The antibody molecules generally have the shape of a letter Y with unique shapes at the end of the branches of the Y that can bind to the antigen. Figure 8.8 illustrates a hypothetical structure of the antibody–antigen structural relationship where the surfaces of the antigen are ball-shaped molecules and the ends of the antibody designated by Y can bond to the antigen as shown in Figure 8.8a and b. Magnetic nanoparticles such as Fe_2O_3 and Fe_3O_4 coated with an antibody specific to one bacteria can be the basis of a rapid detection system of toxic bacteria. There have been a number of demonstrations of the feasibility of such a detection system. For example, it has been demonstrated that NMR can be used to detect bacteria bonded to magnetic nanoparticles [6, 7]. The bacteria that is the cause of tuberculosis has been detected by this method. The demonstration was done using a simulant for this

(a)

(b)

(c)

(d)

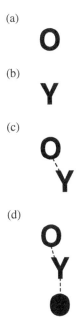

FIGURE 8.8 Representation of an antigen (a), an antibody (b), an antibody bonded to an antigen (c), and a magnetic particle bonded to an antibody and an antigen (d).

bacteria called Bacillus Calmette–Guérin (BCG). Superparamagnetic nanoparticles of Fe_3O_4 were coated with monoclonal antibodies of BCG denoted as SPNBCG. Controlled amounts of both BCG and SPNBCG were added to a fixed amount of human sputa. The bacteria bonded to the particles were detected by NMR. As few as 20 colony-forming units (CFU) were detected in less than 30 min in 1 ml of sputa. The CFU is a unit used in microbiology that measures the number of living bacteria. Generally, it is given as CFU per milliliter for liquids. The principle of the detection is that the inhomogeneous magnetic field produced by the magnetic nanoparticles shortens the spin–spin relaxation time of the protons of the water molecules surrounding the bacteria.

A special NMR microcoil was designed to detect the bacteria. The system, illustrated in Figure 8.9, consists of a microfluidic channel through which the fluid containing the bacteria flows to a membrane, which acts as a filter to material entering the coil. The diameter of the coil is less than 5 mm. The filter separates out magnetic particles that did not bond to bacteria from those that did. Only the particles bonded to the bacteria enter the coil. The change in the spin–spin relaxation time indicates the presence of the bacteria. Figure 8.10 shows a plot of the percent change of the spin–spin relaxation time versus the CFU.

NMR is not the only method that has been used to detect magnetic labeled bacteria or biomolecules. Magnetic tunnel junctions discussed in Chapter 3 have

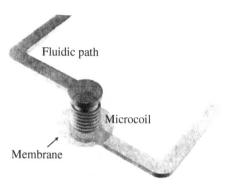

FIGURE 8.9 NMR microcoil, membrane filter, and fluidic path used for detecting TB bacteria (Reproduced with permission from Ref. [6]. © 2009 John Wiley & Sons, Inc.)

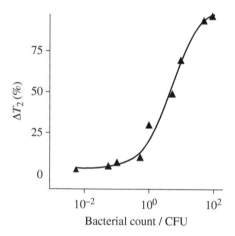

FIGURE 8.10 Plot of the percent change of the spin–spin relaxation time versus number of TB bacteria (Reproduced with permission from Ref. [6]. © 2009 John Wiley & Sons, Inc).

been used to detect the small magnetic field arising from magnetic labeled bio-molecules such as DNA [8].

Some pathogenic bacteria actually consume iron atoms. Hemoglobin (Hb) is a major component of red blood cells whose function is to transport oxygen. Hb whose structure is illustrated in Figure 8.11a consists of folded chains of protein molecules that are made of chains of amino acids. The Hb molecule is approximately spherical having a diameter of 5.5 nm and therefore can be classified as a nanoparticle. Also contained within this structure is the heme molecule, which contains the iron atoms. The structure of the heme molecule is illustrated in Figure 8.11b. The malaria parasite when it is in the blood consumes Hb, but it cannot digest or excrete the iron. Thus, the malaria parasite in the bloodstream contains crystals of iron called hemozoin. Its presence in the blood means it can

(a)

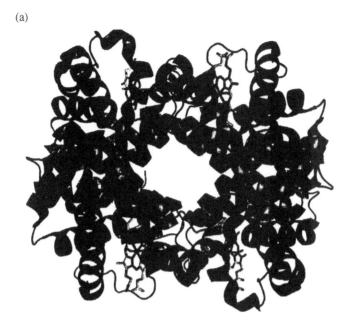

(b)

FIGURE 8.11   Illustration of the structure of hemoglobin (a) and the heme molecule (b).

be detected by magnetic sensors such as SQUID magnetometry or EPR. Magneto-optical effects have also been demonstrated to detect the presence of malaria in the bloodstream [9]. In this method, polarized laser light is incident on a transparent container containing the malaria-contaminated blood. When a paramagnetic hemozoin is present it aligns with the direction of a dc magnetic field applied to the sample perpendicular to the direction of the light beam. This causes a decrease in the transmitted light intensity, which is proportional to the concentration of hemozoin present.

8.6 ANALYSIS OF STORED BLOOD

Blood stored for use in transfusions deteriorates in time. Use of such blood in transfusions can cause serious harm to patients such as organ failure. It is therefore important to insure the blood has not deteriorated before a transfusion. When deterioration

(a)

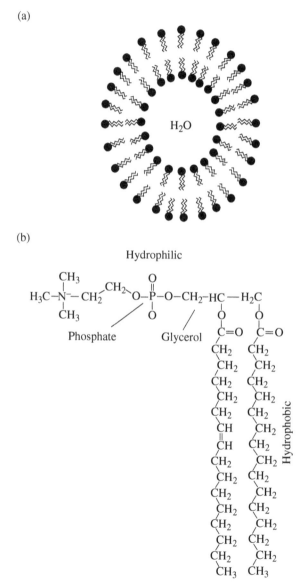

(b)

FIGURE 8.12   (a) Structure of a vesicle consisting of phospholipid molecules surrounding water. (b) A structure of a phospholipid molecule.

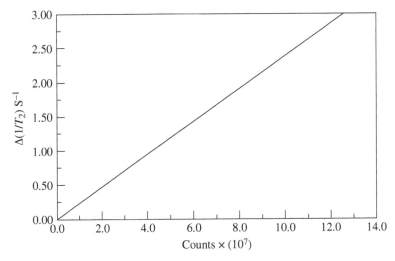

FIGURE 8.13   Plot of the change in the NMR relaxation time versus the number of magnetic nanoparticle-labeled microvesicles (Reprinted with permission from Ref. [10]. © 2013 American Chemical Society).

occurs, the red blood cells, called erythrocytes, produce an increase in microvesicles (MVs) in the blood. A vesicle is an approximately spherical structure having an aqueous interior surrounded by phospholipid molecules as illustrated in Figure 8.12a. Figure 8.12b shows the structure of a phospholipids molecule. One end of the molecule is hydrophilic, meaning it attracts water. The other end is hydrophobic, lacking affinity for water.

The most common method of analyzing stored blood to detect the presence of MVs is flow cytometry. In this method, a stream of liquid is passed by an optical sensor. Laser light is incident on the fluid stream and the scattered light is measured. The technique underestimates the number of MVs in the blood because the MVs are generally smaller than the wavelength of light, which results in weak scattering.

A more accurate method that labels the MVs with magnetic nanoparticles and detects them by NMR measurements of relaxation times has been demonstrated [10]. The MVs are captured on antibody-coated polymer microbeads. The antibodies were further modified by adding chemicals to them such as tetrazine (TZ) and trans-cyclooctene (TCO), which strengthen the bonds between magnetic nanoparticles and the MVs. Iron oxide magnetic nanoparticles are then introduced, which bond to MVs. The number of magnetic nanoparticles attached to the MVs is measured using the NMR transverse relaxation time. Figure 8.13 shows a measurement of the change in inverse of the relaxation time versus the number of MVs counted. The advantages of this method over flow cytometry are increased accuracy, shorter time for analysis, and the need for much less sample typically less than 200 μl.

EXERCISES

8.1 Calculate the force on a 5.8 nm iron nanoparticle in a dc magnetic field that varies in strength as 100 G/cm.

8.2 Why does introducing magnetic nanoparticles into a tumor increase the contrast in the MRI image between the tumor and the normal tissue?

8.3 A 10 nm Fe_2O_3 nanoparticle is suspended in water and in blood. Which would have the larger relaxation time? Explain the reason for your answer.

8.4 A 100 KHz ac magnetic field having an amplitude of 50 G is applied to a fluid containing 10 nm Fe_2O_3 nanoparticles. If the oscillation of the nanoparticles were in phase with the AC field, how much heat would be generated? Give the reason for your answer.

8.5 Calculate the frequency at which an NMR absorption of a hydrogen atom would occur in a one tesla dc magnetic field. What would be the frequency for an O^{16} atom?

REFERENCES

1. J. Frenkel, *Kinetic Theory of Liquids*, Dover Publications, New York, 1955.
2. R. E. Rosenweig, *J. Mag. Mag. Mater.* 252, 370 (2002).
3. P. C. Fannin and S. W. Charles, *J. Phys.* D24, 76 (1991).
4. M. Hanson, *J. Mag. Mag. Mater.* 122, 159 (1991).
5. P. A. Liberti, et al. *J. Mag. Mag. Mater.* 225, 301 (2001).
6. H. Lee, T. Yoon and R. Weissleder, *Angew. Chem. Int. Ed. Engl.* 48, 5657 (2009).
7. C. Kaittanis, S. Naser and J. Manuel Perez, *Nano Lett.* 7, 380 (2007).
8. Y. Shen, B. D. Schrag, M. J. Carterand and G. Xao, *Appl. Phys. Lett.* 93, 033903 (2008).
9. S. Karl, et al. *Malaria J.* 7, 66 (2009).
10. J. Rho, et al. *ACS Nano* 7, 11227 (2013).

9

FABRICATION OF MAGNETIC NANOSTRUCTURES

In this chapter, some examples of methods of fabrication of magnetic nanostructures are presented.

9.1 MAGNETIC NANOPARTICLES

Chemical methods are the most common way to make magnetic nanoparticles.

For example, large-scale chemical synthesis methods of Fe_2O_3 magnetic nanoparticles have been developed. Iron chloride, $FeCl_3 \cdot 6H_2O$, and sodium oleate, $Na[CH_3(CH_2)_7CH=CH(CH_2)_7CO_2]$, are reacted to form an iron oleate complex and NaCl [1]. The iron oleate complex is then slowly heated in 1-octadecene $[CH_3(CH_2)_{15}CH=CH_2]$ to 320°C and held at that temperature for 30 min. The size of the particles can be controlled by the temperature of heating and the time held at the elevated temperature. Figure 9.1 shows a high-resolution transmission electron microscope (TEM) image of different size particles that can be made. It is interesting that the particles are uniform in size and spherical in shape.

Another example is the synthesis of nanosized copper ferrites such as $CuFe_2O_4$ [2]. In this method, oxygen was removed from water by bubbling nitrogen gas through it. Then $Cu(NO_3)_2$ and $Fe(NO_3)_3$ were dissolved in the water in the molar ratio of 1:2. The solution was stirred vigorously and then heated to 70°C for 2 h under a nitrogen atmosphere. Two molar NaOH was slowly added to the solution until a pH of 11 was achieved. The magnetic crystals precipitate out of the solution and are separated from the solution using a magnet. The $CuFe_2O_4$ crystals were then heated for two hours at 400°C to complete the formation of the nanosized crystals. TEM measurements indicated the particle size ranged from 10 to 30 nm.

Physics of Magnetic Nanostructures, First Edition. Frank J. Owens.
© 2015 John Wiley & Sons, Inc. Published 2015 by John Wiley & Sons, Inc.

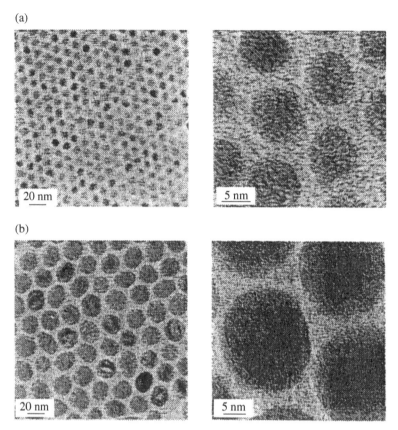

FIGURE 9.1   The particle size in (a) is achieved by a longer heating time compared to the particles in (b) (Adapted from Ref. [1]).

Sol-gels have been utilized to make magnetic nanoparticles. A gel is a semirigid colloidal dispersion of solid particles in a liquid. The gel has a rigidity such that it will not flow under the influence of gravity. Magnetic nanoparticle composites of Fe_3O_4–ZnO have been fabricated using sol-gels [3]. Ethanol (50 ml) was added to a mixture of 3.93 g of zinc acetate and 0.49 g of Fe_3O_4.

A solution of 1.58 ml of diethanolamine and 5 ml of ethanol was slowly added to the mixture until a dark-brown sol was formed. Two hours of stirring converted the sol into a gel. After drying in vacuum at 90°C, the gel was annealed above 350°C for 2 h under vacuum. Figure 9.2 shows a TEM image of the Fe_3O_4–ZnO nanocomposites. The particles are nearly spherical and range in size from 20 to 40 nm. The nanoparticles annealed at 450°C are ferromagnetic having a saturation magnetization of 21 emu/g at 5000 Oe.

Methods other than wet chemical processes have been used to fabricate magnetic nanoparticles. Magnetic nanoparticles can be fabricated by thermal decomposition of solid materials such as ferrocene, $Fe(C_5H_5)_2$ [4]. A mixture of ferrocene and oxalic

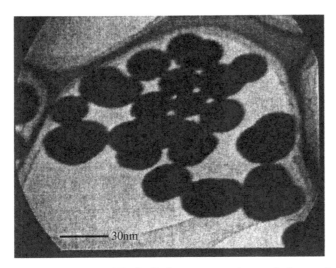

FIGURE 9.2   A TEM image of Fe_2O_3–ZnO nanocomposites made by a sol-gel process (Adapted from Ref. [3]).

acid $[(COOH)_2 \cdot 2H_2O]$ in the ratio by weight of 1:5 to 1 was heated in a ceramic boat in an oven at 1025 K for 10 h. Analysis of the material after decomposition showed the presence of 40 nm ferrite nanoparticles.

Fe_3O_4 can be synthesized by thermal decomposition of iron acetylacetonate, $Fe(C_5H_7O_2)_3$, Fe(Ac), in a solution of stearic acid [5]. The Fe(Ac) is added to the molten stearic acid at 70°C and slowly heated to 180°C until the material turned black. The material was then washed in an organic solution and then dried. This method produced Fe_3O_4 nanoparticles ranging in size from 2 to 12 nm.

9.2   MAGNETIC QUANTUM WELLS

As discussed in Chapter 4, magnetic quantum wells are magnetic materials in which one dimension is of the order of a nanometer. In effect, they are very thin films and can be fabricated by methods used to make thin films. The Au/Fe magnetic quantum wells deposited on the surface of crystals of gallium arsenide, discussed in Section 4.2.1, were fabricated by molecular beam epitaxy (MBE). Figure 9.3 shows a schematic diagram of an MBE apparatus. The system consists of a chamber, which is pumped down to a very high vacuum. The diagram shows three evaporation cells, which are the source of the vapors to be deposited. Typically, the materials are heated in a boron nitride crucible at the base of the evaporation cells to produce the vapors. The vapor beams are interrupted by shutters mounted above the cells, which rotate on a shaft allowing an abrupt change of the beam material and thus precise control of the layer thickness. The substrate on which the layers are deposited is on a molybdenum heating block and is held to the block by indium. The temperature of the substrate is precisely controlled.

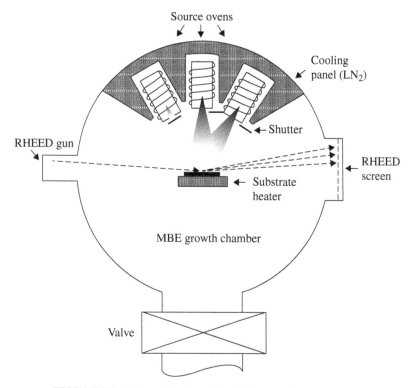

FIGURE 9.3   Schematic of a molecular beam epitaxy apparatus.

The electron gun shown in the figure allows an *in situ* monitoring of the layer-by-layer growth by probing the surface by reflected high-energy electrons. The technique is called reflection high-energy electron diffraction (RHEED). An electron gun produces a beam of high-energy electrons, which are incident on the substrate surface at small angles with respect to the surface. The fluorescent screen shown in the figure detects the diffracted beams.

In Section 4.2.1, the properties of Ni quantum wells were discussed. These wells were fabricated by electrodeposition of Ni on a copper substrate. Figure 9.4 shows an illustration of the apparatus for electrodeposition. It consists of two metal electrodes immersed in a solution and having a voltage applied between them. To make a Ni film, a water solution of $Ni(OH)_2$ is used. The negative electrode, the cathode, is a flat strip of copper. The Ni^{2+} ions in the solution are attracted to the cathode and coat it. The thickness of the film on the copper electrode is determined by the magnitude of the voltage, the concentration of the solute in the solvent, and the length of time the voltage is on.

Thin films of silicon doped with manganese ($Si_{0.95}Mn_{0.05}$) were made using a physical vapor deposition (PVD) process employing sputtering [6]. A schematic of the apparatus is shown in Figure 9.5. In this method, the target in the form of a pellet of Si:Mn is bombarded by positive of ions such as Ar^+, which are accelerated to the

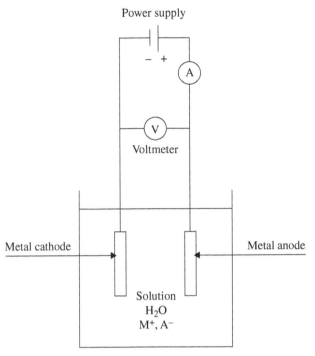

FIGURE 9.4 Illustration of a chemical cell used to fabricate nanoparticles by electrodeposition.

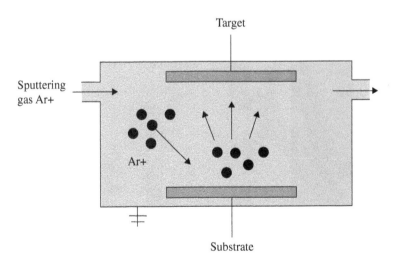

FIGURE 9.5 Schematic of a sputtering apparatus used to fabricate quantum wells.

target by an electric field. The collisions with the target cause atoms of the target to be ejected. These are then collected on the (001) surface of a silicon substrate. The wells made in this manner displayed ferromagnetism at room temperature as discussed in Chapter 4.

Pulsed lasers have also been used to vaporize targets and condense the vapor on substrates. The method, referred to as laser ablation, has been used to create films of $ZnO._{96}CoO._{04}S$ [7].

9.3 MAGNETIC NANOWIRES

Nanowire having diameters of 50–200 nm were fabricated using an electrodeposition process with an aluminum template containing nanosized pores on one of the electrodes [8]. The cell used for the deposition is similar to the one illustrated in Figure 9.4 except that one of the electrodes has an aluminum template on it which is a layer of aluminum having circular nanosized holes in it. The deposition was carried out in a solution of $FeCl_3 \cdot 6H_2O$ and $(NH_4)_2 C_2O_4 \cdot H_2O$. A 70 Hz 15 V ac signal was applied across the electrodes. After completion of the process, the pores of the template contained FeOOH nanowires. Heating the template to 500°C for 3 h followed by reducing the material in flowing H_2 at 325°C produced Fe_2O_3 nanowires. The diameter of the wires could be controlled by the voltage used. Figure 9.6a shows a TEM image of the nanowires having a 150 nm diameter.

Bulk copper oxide undergoes a transition to an antiferromagnetic phase at 230 K. However, as discussed in Chapter 4, when CuO has nanodimensions, it displays ferromagnetism. Nanowires of CuO have been synthesized and are ferromagnetic at room temperature [9]. Equal amounts of copper sulfate ($CuSO_4 \cdot 5H_2O$) and sodium

(a)   (b)

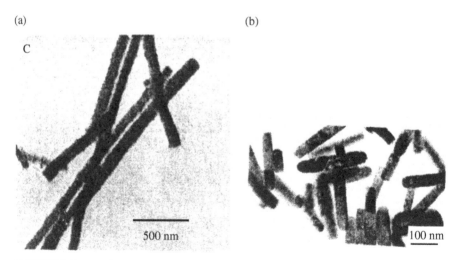

FIGURE 9.6   (a) TEM image of Fe_2O_3 nanowires made by an electrodeposition process and (b) TEM image of CuO nanowires (Adapted from Refs. [8] and [9]).

citrate [Na$_3$(C$_5$H$_5$O$_7$·2H$_2$O)] were dissolved in distilled water and stirred for some time. Then NaOH at about 4 times the amount of the other solvents was added to the solution. The solution was then heated from 100 to 180°C for 12 h in a Teflon-lined autoclave. The separated material was then washed with deionized water and dried at 50°C. Figure 9.6b shows a TEM image of the CuO nanorods made at 180°C.

9.4   MAGNETIC QUANTUM DOTS

A magnetic quantum dot is a ferromagnetic material in which all three dimensions are of nanometer length. Such dots can be fabricated by a top-down process using electron beam lithography. Referring to Figure 9.7, the initial step is to fabricate a quantum magnetic well such as Si:Mn on a substrate. As shown in Figure 9.6a, this is then covered by a material called a resist, which is sensitive to radiation. A typical resist consists of the polymer polymethyl methacrylate [C$_5$O$_2$H$_8$]$_n$. The sample is then irradiated by an electron beam in the region where the nanostructure will be located, as shown in Figure 9.7b. This can be done either by using a radiation mask that contains the nanostructured pattern, as shown. The radiation chemically modifies the exposed area of the resist so that it becomes soluble in a developer. The third step (Fig. 9.7c) in the process is the application of the developer to remove the irradiated portions of the resist. The fourth step (Fig. 9.7d) is the insertion of an etching mask into the hole in the resist, and the fifth step (Fig. 9.7e) consists in lifting off the remaining parts of the resist. In step six (Fig. 9.7f), the areas of the quantum well not covered by the etching mask are chemically etched away to produce the quantum structure shown in Figure 9.7f covered by the etching mask. Finally, the etching mask

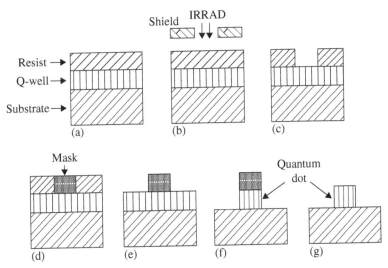

FIGURE 9.7   Illustration of steps in fabrication of magnetic quantum dots by electron beam lithography.

is removed to provide the desired quantum structure (Fig. 9.7g), which is the quantum dot. This is the most common process called electron beam lithography that makes use of an electron beam for the radiation. There are other types of lithography that employ neutral atom beams (e.g., Li, Na, K, Rb, Cs), charged ion beams (e.g., Ga^+), or electromagnetic radiation such as visible light, ultraviolet light, or x-rays.

EXERCISES

9.1 Thermal decomposition of materials can be used to produce magnetic nanoparticles. Suggest some materials that could be used to produce Ni or Co nanoparticles.

9.2 Suggest a method to use electrodeposition to make quantum dots.

9.3 How can LiN_3 be used to make Li nanoparticles?

9.4 Ball milling, grinding, can be used to make small particles. Is this a useful method for making small magnetic nanoparticles?

REFERENCES

1. J. Park, et al. *Nature Mater.* 3, 891 (2006).
2. M. Siddque and N. M. Butt, *Physica* B405, 4211 (2010).
3. A. Hasanpour, M. Niyaifar, M. Asan and J. Amighan, *J. Mag. Mag. Mater.* 334, 41 (2013).
4. A. Bhattacharjee, et al. *J. Mater. Sci.* 48, 2961 (2013).
5. D. K. Jha, et al. *Mater. Lett.* 95, 186 (2013).
6. F. M. Zhang, et al. *Appl. Phys. Lett.* 85, 786 (2004).
7. M. Jadraque, et al. *J. Phys. Chem.* C117, 5416 (2013).
8. L. Zhang and Y. Zhang, *Mag. Mag. Mater.* 321, L15 (2009).
9. H. Xiao, L. Zhu, X. Liu and S. Fu, *Solid State Commun.* 141, 431 (2007).

APPENDIX A

A TABLE OF NUMBER OF ATOMS VERSUS SIZE IN FACE CENTERED CUBIC NANOPARTICLES

Diameter, total number of atoms, number on the surface, and percentage on the surface of rare gas or metallic nanoparticles with face-centered cubic (fcc) close-packed structures. The diameters d in nanometers for some representative fcc solids are Ag 0.289, Al 0.286, Ar 0.372, Au 0.288, Cu 0.256, Fe 0.254, Kr 0.404, Pb 0.350, and Pd 0.275.

Shell	Diameter	Total number	Number on surface	Percentage on surface
1	d	1	1	100
2	$3d$	13	12	92.3
3	$5d$	55	42	76.4
4	$7d$	147	92	62.6
5	$9d$	309	162	52.4
6	$11d$	561	252	44.9
7	$13d$	923	362	39.2
8	$13d$	1415	492	34.8
9	$17d$	2057	642	31.2
10	$19d$	2869	812	28.3
11	$21d$	3871	1002	25.9
12	$23d$	5083	1212	23.8
25	$49d$	4.90×10^4	5.76×10^3	11.7
50	$99d$	4.04×10^5	2.40×10^4	5.9
75	$149d$	1.38×10^6	5.48×10^4	4.0
100	$199d$	3.28×10^6	9.80×10^4	3.0

Physics of Magnetic Nanostructures, First Edition. Frank J. Owens.
© 2015 John Wiley & Sons, Inc. Published 2015 by John Wiley & Sons, Inc.

APPENDIX B

DEFINITION OF A MAGNETIC FIELD

In 1820, the Danish scientist Hans Christian Oersted observed that a magnetic field existed outside a wire carrying a current. Magnetism arises from the forces between electrical charges in motion in contrast to electrostatic forces, which do not require the motion of the charges. Since the electrons in atoms revolve about the nucleus, atoms in certain configurations can display magnetic effects. Instead of dealing with the force exerted on a moving charge by another, the concept of a magnetic field is employed. In this view, a moving charge has a magnetic field in the space around it. This field then exerts a force on a charge moving through it. The symbol $\mathbf{B}$ is used to designate the magnetic induction, which can be depicted by lines of force about the moving charge. The number of lines intersecting a cross-sectional area perpendicular to the lines measures the strength of the vector $\mathbf{B}$.

The definition of $\mathbf{B}$ is as follows. If a positive test charge, Q, at point P moving with a velocity $\mathbf{v}$ experiences a force $\mathbf{F}$, a magnetic induction, $\mathbf{B}$, is present. The relationship between the force and the magnetic field $\mathbf{B}$ is

$$\mathbf{F} = Q\mathbf{v} \times \mathbf{B} \qquad (B.1)$$

where $\mathbf{F}$, $\mathbf{v}$, and $\mathbf{B}$ are vectors. The units of $\mathbf{B}$ are Newtons/(Coulomb-meter/second), which is defined as a weber/meter^2, also called a tesla. Another unit commonly used is the gauss where 10^4 gauss equals one tesla.

If a magnetic field is applied to a paramagnetic material, that is, a material in which the constituents have a net spin but are not ferromagnetically ordered, the magnetic dipoles can align with the direction of the applied field. The sample has a magnetization, $\mathbf{M}$, which is the sum of all the aligned magnetic moments per unit volume. Thus, there is an increase in the magnetic field given by

$$\mathbf{B} = \mu\mathbf{H} + \mathbf{M} \qquad (B.2)$$

where $\mathbf{H}$ is the magnetic field intensity and has the units in the CGS system of oersteds. One oersted is 10^{-4} tesla.

Physics of Magnetic Nanostructures, First Edition. Frank J. Owens.
© 2015 John Wiley & Sons, Inc. Published 2015 by John Wiley & Sons, Inc.

APPENDIX C

DENSITY FUNCTIONAL THEORY

The object of molecular orbital theory is to obtain the energy levels and molecular orbitals (wave functions) and other properties such as the vibrational frequencies. This requires solving the Schrödinger wave equation for a system of many atoms and electrons, which is given by

$$\left[-\sum \frac{(h/\pi)^2}{2m}\left(\nabla_i^2\right) + \sum_{i<j}\frac{e^2}{r_{ij}} - \sum_{iA}\frac{Z_A e^2}{r_{Ai}} + \sum_{A<B}\frac{Z_A Z_B e^2}{R_{AB}} \right]\psi_e = E\psi_e \qquad (C.1)$$

The first term is the kinetic energy where ∇_i^2 is given by

$$\nabla_i^2 = \frac{d^2}{dx_i^2} + \frac{d^2}{dy_i^2} + \frac{d^2}{dz_i^2} \qquad (C.2)$$

The second term represents the electrostatic repulsion between the electrons. The third term is the electrostatic attraction between the electrons and the nuclei, and the fourth term is the electrostatic repulsion between the nuclei. E is the energy and ψ_e is the wave function. Equation C.1 omits the kinetic energy of the nuclei because the electrons move much more rapidly than the nuclei, which are much heavier. In effect, the motion of the nuclei is ignored. This is referred to as the Born–Oppenheimer approximation. The solution of Equation C.1 for molecule having many electrons and nuclei is a very difficult problem. Density functional theory (DFT) provides a simplification that allows a solution for large molecules, nanostructures, and solids.

DFT is an extension of the Thomas–Fermi model of a many-electron atom. The theory treats the electrons as a gas of free electrons confined to a volume V by a spherically symmetric potential. As discussed in Chapter 1, in the context of the free electron model of metals for the electrons confined to a cubic volume, the Fermi level, the highest filled level for a three-dimensional solid is

$$E_{f} = \left[\frac{(h/\pi)^2}{2m}\right]\left[\frac{3\pi^2 N}{V}\right]^{2/3} \tag{C.3}$$

The depth of the potential at any value of r can be related to the density of electrons for that value of r by assuming that the depth of the potential is such that the energy levels are filled to the top, that is, $E_f = -V(r)$. This yields a relationship between the potential and the density of electrons, $\rho = N/V$, given by

$$-V[r] = \left[\frac{(h/\pi)^2}{2m}\right]\left[3\pi^2\rho\right]^{2/3} \tag{C.4}$$

The model assumes the $V(r)$ does not change significantly in lengths compared to the wavelengths of the electrons. This means a number of electrons can be localized in a volume in which $V(r)$ is constant.

The importance of this result is that it transforms the electrostatic interaction of the electrons with the nuclei and the other electrons (the second and third terms in Equation C.1) to a form where the interaction between each electron and every other electron can be represented by the interaction of the electron and nuclei with a charge density. For a molecule having many nuclei and electrons, the electrostatic interaction of the electrons with each other and the nucleus becomes that of noninteracting electrons in an effective potential, which is a function of the charge density.

In DFT, the ground state energy of a many-electron system is a function only of the electron density $\rho(r)$. The wave function must satisfy a Schrödinger-like equation having the form

$$\left[-\sum\frac{(h/\pi)^2}{2m}\left(\nabla_i^2\right) + V_N[\mathbf{r}] + \int\frac{\rho[\mathbf{r'}]}{\mathbf{r}-\mathbf{r'}}d^3r' + \epsilon\left(\rho[\mathbf{r}]\right)\right]\Psi = E_i\Psi \tag{C.5}$$

All terms except the electrostatic repulsion between the nuclei are a function of the charge density. The first term is the kinetic energy. The kinetic energy density at each point is assumed to correspond to the kinetic energy density of a homogeneous noninteracting electron gas, which is proportional to $\rho^{5/3}$. The second is the interaction of the electrons with the nucleus. The third term is the electrostatic repulsion of the electrons with each other in terms of an electron density. The last terms represent the exchange and correlation interactions. In DFT, accurate formulas for ϵ have been developed from simulations of a uniform noninteracting electron gas. The model used to calculate the exchange interaction in DFT is based on the idea to treat the electronic structure of a simple metal such as lithium as a homogeneous gas of electrons around a lattice of positive charges. This model is used in DFT because it is the only one that yields an exact and accurate form of the exchange interaction. The model is referred to as the local density approximation (LDA), and the exchange term has the form

$$E_{xc}^{LDA}[\rho] = \int \rho[\mathbf{r}] \epsilon_{xc}\left(\rho[\mathbf{r}]\right) dr \qquad (C.6)$$

where $\epsilon_{xc}[\rho(\mathbf{r})]$ is the energy density, that is, the exchange plus correlation energy per electron. In a homogeneous electron gas with electron density ρ, it is assumed that the electron density "ρ" varies slowly with position.

The specific form of the exchange energy was obtained by Slater and is given by

$$\epsilon_{xc} = \left(-\frac{3}{4}\right)\left(\frac{3\rho[\mathbf{r}]}{\pi}\right)^{1/3} \qquad (C.7)$$

Substituting into Equation C.6 gives the following form for the exchange energy:

$$E_{xc}^{LDA}[\rho] = \left(-\frac{3}{4}\right)\left[\frac{3}{\pi}\right]^{1/3} \int \rho[\mathbf{r}]^{4/3} \, dr \qquad (C.8)$$

A brief overview of DFT approach within the LDA approximation has been presented without going into the details and proofs of the Hohenberg–Kohn existence and variational theorems. The basic understanding that the reader should be left with is that DFT transforms the complex many-body problem of interacting electrons in the external potential of the nucleus to a tractable problem of noninteracting electrons moving in an effective potential, which is a function of the electron charge density. With the DFT approach, it became possible to perform calculations on quite large molecules with good to excellent accuracies. The theory has been used to calculate the properties of magnetic nanostructures.

APPENDIX D

TIGHT BINDING MODEL OF ELECTRONIC STRUCTURE OF METALS

The tight-binding model takes a different approach to treating the electronic structure of metals. It assumes that when the electron is close to the atom of the solid, it will have a wave function $\Phi_a(r-l)$ corresponding to the wave function of the free atom. When the electron is far from the atom, the wave function can be described as that of a free electron, that is, $\exp(iK \cdot l)$. The wave function is written in this model as

$$\Psi_k[r] = \sum_l \exp(iK \cdot l)\Phi_a(r-l) \tag{D.1}$$

The function looks like a series of strongly localized atomic orbitals whose amplitude is modulated by a phase factor. The model is appropriate to materials having atoms with d orbitals such as transition metals, which are compact and form narrow well-defined bands. The energy levels are determined by

$$E = E_0 + \left\langle \Psi_k^*[r]H\Psi_k[r] \right\rangle \tag{D.2}$$

which is

$$E = E_0 + \sum_l \exp(iK \cdot l)\left\langle \Phi_a(r-l)H\Phi_a(r) \right\rangle \tag{D.3}$$

where H is the difference between the potential of the free atom and the potential the atom sees in the crystal, $H = V(r) - V_a(r)$. For the case of a monatomic cubic

Physics of Magnetic Nanostructures, First Edition. Frank J. Owens.
© 2015 John Wiley & Sons, Inc. Published 2015 by John Wiley & Sons, Inc.

crystal having lattice parameter, a, and assuming only nearest neighbor interactions, Equation D.3 becomes

$$E = E_0 + 2t(\cos K_x a + \cos K_y a + \cos K_z a) \tag{D.4}$$

where t is given by

$$\langle \Phi_a(r-l)H\Phi_a(r)\rangle \tag{D.5}$$

It turns out that t is negative. $\Phi_a(r)$ for the top occupied band is the highest occupied atomic orbital of the free atom, and E_0 is the energy of the top occupied orbital of the free atom. The lowest energy of the band is $E_0 - 2t$ occurring at $k=0$, and the highest energy occurs at $E_0 + 2t$, which means the bandwidth is $4t$. The separation between the highest and lowest energy depends on t, which involves integrals of the atomic orbitals over the potential in the crystal given by Equation D.5. If as in the case of many metals the lattice parameter decreases with nanosizing (as discussed in Section 1.1), then the separation between the energy at $K=0$ and $K=\pi$ increases because t increases. Consider a one-dimensional finite lattice of length L having N atoms separated from each other by a. The number of states in an energy band will be N corresponding to the allowed wavelengths of the wave function that fit into L such that $\psi_N(0)=0$ and $\psi_N(L)=0$. In the tight-binding model, the dependence of E on k in a band will be

$$E = E_0 - 2t \cos Ka \tag{D.6}$$

where K can have values from 0 to $N\pi/L$. The density of states in the band is approximately $N/2t$. Thus, as in the free electron model, the density of states decreases with the number of atoms in the particle and therefore the size of the particle.

APPENDIX E

PERIODIC BOUNDARY CONDITIONS

The use of periodic boundary conditions is possible because of the Bloch theorem.

The theorem states that for any wave function that satisfies the Schrödinger equation, there exists a wave vector, K, such that a translation by a lattice vector $\mathbf{L}$ is equivalent to multiplying the wave function by a phase factor $\exp(iK\cdot\mathbf{L})$, that is,

$$\Psi_K(r+\mathbf{L}) = \exp(iK\cdot\mathbf{L})\Psi_K(r) \qquad (\text{E.1})$$

Consider a one-dimensional lattice having a lattice vector $\mathbf{L}$. To correctly obtain the energy levels, we would have to consider a lattice of length $N\mathbf{L}$ where N is infinite. This of course is not computationally possible. Now, if N were finite, it would be necessary that the wave function be zero at the ends of the lattice. Because of reflections from the ends, this would lead to standing waves, which would have to be included in the solution and do not exist in large crystals.

The periodic boundary conditions or Born–von Karman boundary conditions provide a mathematical device to get around the physical effects of boundaries. In one dimension, the device forms the lattice into a circle of cells. To insure that there is no discontinuity of the wave function, it is required that

$$\Psi(x+\mathbf{L}a) = \Psi(x) \qquad (\text{E.2})$$

where a is the lattice parameter. The Bloch condition in one dimension is

$$\Psi_K(x+\mathbf{L}a) = \exp(iK\mathbf{L}a)\Psi_K(X) \qquad (\text{E.3})$$

Thus, $\exp(iK\mathbf{L}a)=1$, which means $K=2\pi m/\mathbf{L}a$ where m is an integer. In the reduced zone in one dimension, K must have values as $-\pi/a < K > \pi/a$. The integers m go from $-(1/2)L$ to $(1/2)L$. The number of allowed wave vectors in the Brillouin zone equals the number of unit cells in the crystal.

Physics of Magnetic Nanostructures, First Edition. Frank J. Owens.
© 2015 John Wiley & Sons, Inc. Published 2015 by John Wiley & Sons, Inc.

INDEX

Physics of Magnetic Nanostructures, First Edition. Frank J. Owens.
© 2015 John Wiley & Sons, Inc. Published 2015 by John Wiley & Sons, Inc.